Liberty Hyde Bailey

American Grape Training

An Account of the Leading Forms Now in Use of Training the American

Grapes

Liberty Hyde Bailey

American Grape Training
An Account of the Leading Forms Now in Use of Training the American Grapes

ISBN/EAN: 9783744678681

Printed in Europe, USA, Canada, Australia, Japan

Cover: Foto ©berggeist007 / pixelio.de

More available books at **www.hansebooks.com**

AMERICAN GRAPE TRAINING

An account of the leading forms now in use of Training the American Grapes.

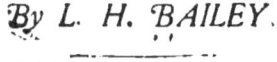
By L. H. BAILEY.

NEW YORK:
THE RURAL PUBLISHING COMPANY
1893.

By the same Author.

Annals of Horticulture in North America for the year 1889. A witness of passing events and a record of progress. 249 pages, 52 illustrations.
Annals for 1890. 312 pages, 82 illustrations.
Annals for 1891. 416 pages, 77 illustrations.
Annals for 1892.
*⁎*A new volume is issued each year, each complete in itself. Cloth, $1; paper, 60 cents.
The Horticulturist's Rule-Book. A compendium of useful information for fruit-growers, truck-gardeners, florists and others. Second edition, revised to the opening of 1892. 221 pages. Cloth, $1; paper, 50 cents.
The Nursery Book. A complete guide to the multiplication and pollination of plants 304 pages, 106 illustrations. Cloth, $1; paper, 50c.
Cross-Breeding and Hybridizing. With a brief bibliography of the subject. 44 pages. Paper, 40 cents. (Rural Library Series.)
Field Notes on Apple Culture. 90 pages, 19 illustrations. Cloth, 75 cents.
Talks Afield: About plants and the science of plants. 173 pages, 100 illustrations. Cloth, $1.

COPYRIGHTED 1893,
BY L. H. BAILEY.

ELECTROTYPED AND PRINTED BY
J. HORACE M'FARLAND CO., HARRISBURG, PA.

CONTENTS.

CHAPTER I.

	Pages
Introduction	9 11
Pruning	11-24

CHAPTER II.

Preliminary Preparations for Training—The Trellis—Tying 25-33

CHAPTER III.

The Upright Systems. (Horizontal Arm Spur System. High Renewal. Fan Training) 34-55

CHAPTER IV.

The Drooping Systems. (True or Four-Cane Kniffin. Modifications of the Four-Cane Kniffin. The Two-Cane Kniffin or Umbrella System. The Low or One-Wire Kniffin. The Six-Cane Kniffin. Overhead, or Arbor Kniffin. The Cross-Wire System. Renewal Kniffin. The Munson System) 56-82

CHAPTER V.

Miscellaneous Systems. (Horizontal Training. Post Training. Arbors. Remodeling Old Vines) . . . 83-92

ILLUSTRATIONS.

		PAGE
1.	Grape Shoot	12
2.	The Bearing Wood	13
3.	Diagram	15
4.	Spur	18
5.	Renewal Pruning	19
6.	A Newly Set Vineyard	21
7.	Horizontal Arm Spur Training	35
8.	Horizontal Arm (Diagram)	36
9.	Short Arm Spur Training	38
10.	The Second Season of Upright Training	40
11.	Making the T-Head	42
12.	The Third Season of High Renewal	43
13.	High Renewal, before Pruning	44
14.	High Renewal, Pruned	45
15.	High Renewal, Pruned and Tied	46
16.	High Renewal with Four Canes	47
17.	High Renewal Complete	48
18.	A Slat Trellis, with Upright Training	51
19.	Fan Training, after Pruning	55
20.	William Kniffin	57
21.	The True Kniffin Training	59
22.	No. 21, when Pruned	60
23.	A Poor Type of Kniffin	64
24.	The Y-Trunk Kniffin	65
25.	Umbrella Training	67
26.	A Poor Umbrella System	68
27.	Eight-Cane Kniffin (Diagram)	70
28.	Overhead Kniffin	71
29.	Overhead Kniffin	72
30.	Overhead Kniffin, before Pruning	73
31.	Cross-Wire Training	75
32.	Cross-Wire Training, Outside View	76
33.	Munson Training. End View	78
34.	Munson Training. Side View	79
35.	Horizontal Training	83
36.	Low Post Training	86
37.	A Yearling Graft	91

PREFACE.

THIS LITTLE book has grown out of an attempt to teach the principles and methods of grape training to college students. I have found such teaching to be exceedingly difficult and unsatisfactory. It is impossible to firmly impress the lessons by mere lectures. The student must apprehend the principles slowly and by his own effort. He must have time to thoroughly assimilate them before he attempts to apply them. I therefore cast about for books which I could put before my class, but I at once found that there are very few succinct accounts of the subjects of grape pruning and training, and that none of our books portray the methods which are most largely practised in the large grape regions of the east. My only recourse, therefore, was to put my own notes into shape for print, and this I have now done. And inasmuch as all grape-growers are students, I hope that the simple account will find a use beyond the classroom.

This lack of adequate accounts of grape training at first astonished me, but is not strange after all. It must be remembered that the cultivation of the native grape is of very recent origin. There are many men who can remember its beginning in a commercial way. It seldom occurs to the younger generation, which is familiar with the great vineyards in many states, that the Concord is yet scarcely forty years old, and that all grape growing in eastern America is yet in an experimental stage. Progress has been so rapid in recent years that the new methods outstrip the books. The old horizontal arm spur system, which is still the chief method in the books, has evolved itself into a high renewal training, which is widely used but which has not found its way into the manuals. The Kniffin type has outgrown its long period of incubation, and is now taking an assured place in vineyard management. So two great types, opposed in method, are now contending for supremacy, and they will probably form the basis of all future developments. This evolution of American grape training is one of the most unique and signal developments of our modern horticulture, and its very recent departure from the early doubts and trials is a fresh illustration of the youth and virility of all horticultural pursuits in North America.

This development of our grape training should form the subject of a historical inquiry. I have not attempted such in this little hand-book. I have omitted all reference to the many early methods, which were in most cases transportations or modifications of European practices, for their value is now chiefly historical and their insertion here would only confuse the reader. I have attempted nothing more than a plain account of the methods now in use; in fact, I am aware that I have not accomplished even this much, for there are various methods which I have not mentioned. But these omitted forms are mostly of local use or adaptation, and they are usually only modifications of the main types here explained. It is impossible to describe all the variations in grape training in a book of pocket size; neither is it necessary. Nearly every grower who has given grape raising careful attention has introduced into his own vineyard some modifications which he thinks are of special value to him. There are various curious and instructive old books to which the reader can go if he desires to know the history and evolution of grape training in America. He will find that we have now passed through the long and costly experiment with European systems. And we have also outgrown the gross or long-wood styles, and now prune close with the expectation of obtaining superior and definite results.

I have not attempted to rely upon my own resources in the preparation of this book. All the manuscript has been read by three persons—by George C. Snow, Penn Yan, N. Y., William D. Barns, Middle Hope, N. Y., and L. C. Corbett, my assistant in the Cornell Experiment Station. Mr. Snow is a grower in the lake region of western New York, and employs the High Renewal system; Mr. Barns is a grower in the Hudson River valley, and practices the Kniffin system ; while Mr. Corbett has been a student of all the systems and has practiced two or three of them in commercial plantations. These persons have made many suggestions of which I have been glad to avail myself, and to them very much of the value of the book is to be attributed.

L. H. BAILEY.

ITHACA, N. Y., *Feb. 1, 1893.*

JOHN ADLUM, of the District of Columbia, appears to have been the first person to systematically undertake the cultivation and amelioration of the native grapes. His method of training, as described in 1823, is as follows: One shoot is allowed to grow the first year, and this is cut back to two buds the first fall. The second year two shoots are allowed to grow, and they are tied to "two stakes fixed down to the side of each plant, about five or six feet high;" in the fall each cane is cut back to three or four buds. In the third spring, these two short canes are spread apart "so as to make an angle of about forty-five degrass with the stem," and are tied to stakes; this season about two shoots are allowed to grow from each branch, making four in all, and in the fall the outside ones are cut back to three or four buds and the inner ones to two. These outside shoots are to bear the fruit the fourth year, and the inside ones give rise to renewal canes. These two outer canes or branches are secured to two stakes set about sixteen inches upon either side of the vine, and the shoots are tied up to the stakes, as they grow. The renewal shoots from the inside stubs are tied to a third stake set near the root of the vine. The outside branches are to be cut away entirely at the end of the fourth year. This is an ingenious renewal post system, and it is easy to see how the Horizontal Arm and High Renewal systems may have sprung from it.

AMERICAN GRAPE TRAINING.

CHAPTER I.

INTRODUCTION—PRUNING.

Pruning and training the grape are perplexed questions, even to those who have spent a lifetime in grape growing. The perplexity arises from several diverse sources, as the early effort to transplant European methods, the fact that many systems present almost equally good results for particular purposes and varieties, and the failure to comprehend the fundamental principles of the operations.

It is sufficient condemnation of European methods when applied in eastern America, to say that the American grapes are distinct species from the European grapes, and that they are consequently different in habit. This fact does not appear to have been apprehended clearly by the early American grape-growers, even after the native varieties had begun to gain prominence. American viticulture, aside from that upon the Pacific slope which is concerned with the European grape, is an industry of very recent development. It was little more than a century ago that the first American variety gained favor, and so late as 1823 that the first definite attempt was made, in Adlum's "Memoir on the Cultivation of the Vine in America," to record the merits of native grapes for purposes of cultivation. Even Adlum's book was largely given to a discussion of European varieties and practices. In 1846 "Thomas' Fruit Culturist" mentioned only six "American hardy varieties," and all of these, save the Catawba, are practically not in cultivation at the present time. The Concord appeared in 1853. American grape training is, therefore, a very recent development, and we are only now outgrowing the influence of the practices early imported from Europe. The first decided epoch in the evolution of our grape training was the appearance of Fuller's "Grape Culturist," in 1864; for while the system which he depicted and which yet often bears his name, was but a modification of some European methods and had been outlined by earlier American writers, it was at that time placed clearly and cogently before the public and became an accepted practice. The fundamental principles of pruning are alike for both European and American grapes, but the details of pruning and training must be greatly modified for different species. We must understand at the outset that American species of grapes demand an American system of treatment.

The great diversity of opinion which exists

amongst the best grape growers concerning the advantages of different systems of training is proof that many systems have merit, and that no one system is better than others for all purposes. The grower must recognize the fact that the most important factor in determining the merits of any system of training is the habit of the vine—as its vigor, rate of growth, normal size, relative size and abundance of leaves, and season and character of fruit. Nearly every variety differs from others in habit in some particular, and it therefore requires different treatment in some important detail. Varieties may thrive equally well upon the same general system of training, but require minor modifications; so it comes that no hard and fast lines can be laid down, either for any system or any variety. One system differs from another in some one main principle or idea, but the modifications of all may meet and blend. If two men practice the Kniffin system, therefore, this fact does not indicate that they prune and train their vines exactly alike. It is impossible to construct rules for grape training; it is, therefore, important that we understand thoroughly the philosophy of pruning and training, both in general and in the different systems which are now most popular. These points we shall now consider.

PRUNING.

Pruning and training are terms which are often confounded when speaking of the grape, but they represent distinct operations. Pruning refers to such removal of branches as shall insure better and larger fruit upon the remaining portions. Training refers to the disposition of the different parts of the vine. It is true that different methods of training demand different styles of pruning, but the modification in pruning is only such as shall adapt it to the external shape and size of the vine, and does not in any way affect the principle upon which it rests. Pruning is a necessity, and, in essence, there is but one method; training is largely a convenience, and there are as many methods as there are fancies among grape growers.

All intelligent pruning of the rape rests upon the fact that *the fruit is borne in a few clusters near the base of the growing shoots of the season, and which spring from wood of last year's growth.* It may be said here that a growing, leafy

1. GRAPE SHOOT.

Pruning.

branch of the grape vine is called a *shoot ;* a ripened shoot is called a *cane ;* a branch or trunk two or more years old is called an *arm.* Fig. 1 is a shoot as it appears in the northern states in June. The whole shoot has grown within a month, from a bud. As it grew, flower clusters appeared and these are to bear the grapes. Flowering is now

2. THE BEARING WOOD.

over, but the shoot will continue to grow, perhaps to the length of ten or twenty feet. At picking time, therefore, the grapes all hang near the lower end or base of the shoots or new canes, as in fig. 2. Each bud upon the old cane, therefore, produces a new cane, which may bear fruit as well as leaves. At the close of the season, this long ripened shoot or cane has produced a bud every foot or less, from which new fruit-bearing shoots are to spring next year. But if all these buds were allowed to remain, the vine would be overtaxed with fruit the coming year and the crop would be a failure. The cane is, therefore, cut off until it bears only as many buds as experience has taught us the vine should carry. The cane may be cut back to five or ten buds, and perhaps some of these buds will be removed, or "rubbed off," next spring if the young growth seems to be too thick, or if the plant is weak. Each shoot will bear, on an average, two or three clusters. Some shoots will bear no clusters. From one to six of the old canes, each bearing from five to ten buds, are left each spring. The number of clusters which a vine can carry well depends upon the variety, the age and size of the vine, the style of the training, and the soil and cultivation. Experience is the only guide. A strong vine of Concord, which is a prolific variety, trained upon any of the ordinary systems and set nine or ten feet apart each way, will usually carry from thirty to sixty clusters. The clusters will weigh from a fourth to a half pound each. Twelve or fifteen pounds of marketable grapes is a fair or average crop for such a Concord vine, and twenty-five pounds is a very heavy crop.

The pruning of the grape vine, therefore, is

Pruning. 15

essentially a thinning process. In the winter pruning, all the canes of the last season's growth are cut away except from two to six, which are left to make the fruit and wood of the next year; and each of these remaining canes is headed back to from three to ten buds. The number and length of the canes which are left after the pruning depend upon the style of training which is practiced. A vine which

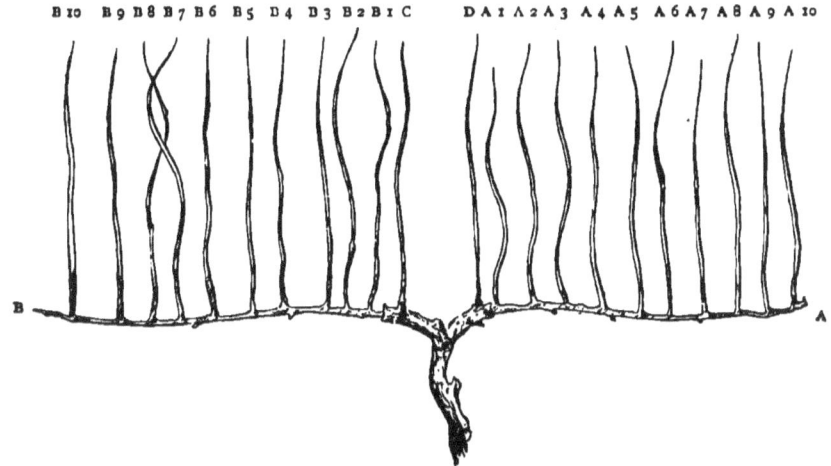

3. DIAGRAM

may completely cover a trellis in the fall, will be cut back so severely that a novice will fear that the plant is ruined. But the operator bears in mind the fact that the grape, unlike the apple, pear and peach, does not bear distinct fruit-buds in the fall, but buds which produce both fruit and wood the following season.

Let us now suppose, therefore, that we have pruned our vine in the fall of 1891 to two canes, each bearing ten buds. We will call these canes A and B, respectively. (Fig. 3.) In 1892, therefore, twenty shoots grow from them, and each of these shoots or new canes branches, or produces laterals. We will call these new canes of 1892, A 1, A 2, A 3, B 1, B 2, and so on. Each of the new canes bears at the base about two clusters of grapes, giving a total yield of about forty clusters. These clusters stand opposite the leaves, as seen in fig. 1. In the axil of each leaf a bud is formed which will produce a cane, and perhaps fruit, in 1893. If each of these new canes, A 1, A 2, etc., produce ten buds—which is a moderate number—the vine would go into the winter of 1892-3 with 200 buds for the next year's growth and crop; but these buds should be reduced to about twenty, as they were in the fall of 1891. That is, every year we go back again to the same number of buds, and the top of the vine gets no larger from year to year. We must, therefore, cut back again to two canes. We cut back each of the original canes, A and B, to one new cane. That is, we leave only A 1 and B 1, cutting off A 2, A 3, etc., and B 2, B 3, etc. This brings the vine back to very nearly its condition in the fall of 1891; but the new canes, A 1 and B 1, which are now to become the main canes by being bent down horizontally, were borne at some distance—say three or four inches—from the base of the original canes, A and B, so that the permanent part of the vine is constantly lengthening itself.

This annually lengthening portion is called a *spur*. Spurs are rarely or never made in this exact position, however, although this diagrammatic sketch illustrates clearly the method of their formation. The common method of spurring is that connected with the horizontal arm system of training, in which the canes A and B are allowed to become permanent arms, and the upright canes, A 1, A 2, B 1, B 2, B 3, etc., are cut back to within two or three buds of the arms each year. The cane A 1, for instance, is cut back in the fall of 1892 to two or three buds, and in 1893 two or three canes will grow from this stub. In the fall of 1893 only one cane is left after the pruning, and this one is cut back to two or three buds; and so on. So the spur grows higher every year, although every effort is made to keep it short, both by reducing the number of buds to one or two and by endeavoring to bring out a cane lower down on the spur every few years. Fig. 4 shows a short spur of two years' standing. The horizontal portion shows the permanent arm. The first upright portion is the remains of the first-year cane and the upper portion is the second-year cane after it is cut back in the fall. In this instance, the cane is cut back to one fruiting bud, *b*, the small buds, *a a*, being rubbed out. There are serious objections to spurs in any position. They become hard and comparatively lifeless after a time, it is often difficult to replace them by healthy fresh wood, and the bearing portion of the vine is constantly receding from the main trunk. The bearing wood should spring from near the central portions of the vine, or be kept "near the head," as the grape-growers say. In order to do this, it is customary to allow two canes to grow out each year back of the canes A 1 and B 1, or from the head of the vine ; these canes may be designated C and D. (Fig. 3.) These canes, C and D. are grown during 1892— when they may bear fruit like other canes—for the sole purpose of forming the basis of the bearing top in 1893, while all the old top, A and B, with the secondary canes, A 1, A 2, B 1, B 2, B 3, etc., is cut entirely away. Here, then, are two distinct methods of forming the bearing top for the succeeding year : either from *spurs*, which are the remains of the previous top ; or from *renewals*, which are taken each year from the old wood near the head of the vine, or even from the ground. Renewals from the ground are now little used, however, for they seldom give a sufficient crop unless they are headed in the first fall and are allowed to bear the second year. It should be borne in mind that the spur and renewal methods refer entirely to pruning, not to training, for either one can be used

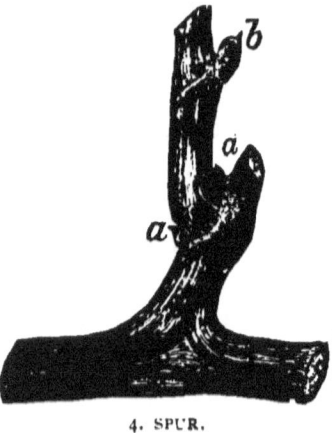
4. SPUR.

Pruning.

in any system of training. Spur pruning, however, is growing in disavor amongst commercial grape-growers, and the renewal is more or less used in all systems of training.

Fig. 5 illustrates a renewal pruning. This engraving shows the head of a vine seven years old, and upon which two canes are allowed to remain after each annual pruning. The portion extending

5. RENEWAL PRUNING.

from *b* to *f* and *d* is the base of the bearing cane of 1892. In the winter of 1892-3, this cane is cut off at *d*, and the new cane, *e*, is left to make the bearing wood of 1893. Another cane sprung from *f*, but it was too weak to leave for fruiting. It was, therefore, cut away. The old stub, *b, f, d*, will be cut away a year hence, in the winter of 1893-4. In the meantime, a renewal cane will have grown from the stub *c*, which is left for that purpose, and the old cane, *b d*, will be cut off just beyond it, between *c* and *f*. In this way, the bearing wood is kept close to the head of the vine. The wound *a* shows where an old stub was cut away this winter, 1892-3, while *b* shows where one was cut off the previous winter. A scar upon the back of the head, which does not show in the illustration, marks the spot where a stub was cut away two years ago, in the winter of 1890-1. This method of pruning can be kept up almost indefinitely, and if care is exercised in keeping the stubs short, the head will not enlarge out of proportion to the growth of the stock or trunk.

Pruning Young Vines.—The time required after planting to get the vine onto the wires or trellis varies with the strength of the vine when set, the variety, the soil and cultivation, and the system of training; but, as a rule, the training begins the second or third year, previous to which time the vine is pruned, not trained. Two-year-old vines are most popular for planting, although in the strong varieties, like Concord and Niagara, well-grown yearling vines are probably as good, if not better. The strong-growing kinds are commonly set from eight to ten feet apart in the row, and the rows eight or nine feet apart. Delawares and other small vines may be set closer, although eight feet is preferable. When set, the vine is cut back to two or three buds. During the first year, the young canes are usually allowed to lie upon the ground at will, as seen in fig. 6. In the fall or winter, all the canes but one are cut off, and this one is cut back to two or three buds. The vine is, therefore, no larger at the expiration of a year's growth than it was when planted; but in the meantime the plant has become thoroughly established in the soil, and the second year's growth should be strong enough to form the basis for the

6. A NEWLY SET VINEYARD.

permanent trunk or arm. If, however, the second year's growth is weak, it may be cut back as before, and the third season's growth used for the trunk. On the other hand, the growth of the first year is sometimes carried onto the wires to form the permanent trunk and arms, but it is only with extra strong vines in good soil that this practice is admissible. From this point, the treatment of the vine is discussed under training.

When to Prune.—Grape vines may be pruned at any time during the winter. It is the practice among most grape-growers in the north to prune as time permits from November to late in February, or even early March. The sap flows very freely from cuts made in spring and early summer, causing the phenomenon known as "bleeding," or in Europe as "weeping," and in order to prevent this loss, pruning is stopped six weeks or more before the time at which the buds usually swell. It is yet a moot point if this bleeding injures the vine, but it is a safe practice to prune early. The vine is cut off an inch or two beyond the last bud which it is desired to leave, in order to avoid injury to the bud from the drying out of the end of the cane.

The pruning is done with small hand pruning-shears. The canes are often allowed to remain tied to the wires until the pruning is accomplished, although it is the practice with most growers who use the Kniffin system to cut the strings before pruning. The removal of the severed canes is known as "stripping." In large vineyards, the pruner sometimes leaves the stripping to boys or other cheap labor. The stripping may be done at any time after the pruning is performed until spring. It must be done before the growth starts on the remaining portions of the vine, however, to avoid injury to the young buds when tearing the vines off the trellis.

Summer Pruning.—There is much discussion as to the advisability of summer pruning. It is essential to the understanding of the question that the grower bear in mind that this summer pruning is of two kinds—the removal or "breaking out" of the superfluous shoots, and heading-in or "stopping" the main canes to keep them within limits. The superfluous shoots are such as spring from small, weak buds or those which break from the old arms or trunk of the vine. Shoots which start from the very base of the old cane are usually weak and should be removed. Buds in this position are shown at *a a*, in fig. 4. The secondary or axillary branches, which often start from the base of the season's shoots, should be removed or broken out. These superfluous shoots are pulled off from time to time as they appear, or the buds may be rubbed off before the shoots begin to grow.

The heading-in of the main canes, while desirable for the purpose of keeping the vine within bounds, is apt to cause a growth of laterals which choke up the vine and which do not mature, and in those styles of training in which very little wood is allowed to grow, the practice may prevent the development of a sufficient amount of leaf surface to properly sustain the vine. Vines are often weakened by summer pruning. These dangers can be overcome by careful attention, however, especially by heading-in very lightly and by doing it as late in the season as possible, when new lateral growth does not start readily. The necessity of much heading-in has been largely obviated in late years by the adoption of high or drooping systems of training, and by setting the vines far apart. The strong varieties, like Concord, Brighton and Niagara, should be set ten feet apart in the row, especially if grown upon the Kniffin system. Catawba, being a very upright grower and especially well adapted to upright training, may be set eight feet apart, and Delawares are often set as close as six or eight feet. It is doubtful, however, if any variety should be set less than eight feet apart for trellis culture. In Virginia and southward, where the growth is large because of the long seasons, vines are often set more than ten feet apart. In the South, the rows should run north and south, that the fruit may be shaded from midday sun. The only summer heading-in now generally recommended is the clipping of the tips when they fall over and begin to touch the ground. This clipping is often done with a sickle or sharp corn-cutter.

Objects of Pruning.—The objects of pruning the grape, as of other fruits, are five :

1. To produce larger and better fruit.
2. To maintain or augment the vigor of the vine.
3. To keep the vine within manageable limits.
4. To facilitate cultivation.
5. To facilitate spraying.

CHAPTER II.

PRELIMINARY PREPARATIONS FOR TRAINING — THE TRELLIS—TYING.

Training the grape vine is practiced for the purpose of keeping the vine in convenient shape and to allow each cluster to receive its full amount of space and light. A well trained vine is easily cultivated and sprayed, and the grapes are readily harvested, and it is only upon such vines that the best and fairest fruit is uniformly produced. Some kind of training is essential, for a vine will not often bear good fruit when it lies upon the ground. In essence, there are three general types or styles of training, which may be designated as the upright, drooping and horizontal, these terms designating the direction of the bearing shoots. The upright systems carry two or more canes or arms along a low horizontal wire, or sometimes obliquely across a trellis from below upwards, and the shoots are tied up as they grow to the wires above. The horizontal systems carry up a perpendicular cane or arm, or sometimes two or more, from which the shoots are carried out horizontally and are tied to perpendicular wires or posts. The drooping systems, represented in the Kniffin and post-training, carry the canes or arms upon a high horizontal wire or trellis and allow the shoots to hang without tying. To one or another of these types all the systems of American grape-training can be referred.

There is no system of training which is best for all purposes and all varieties. The strong-growing varieties more readily adapt themselves to the high drooping systems than the weaker varieties, although the Delaware is often trained on a comparatively low Kniffin with good effect. The high or drooping systems are of comparatively recent date, and their particular advantages are the saving of labor in summer tying, cheapness of the trellis, and the facility with which the ground can be cultivated without endangering the branches of the vine. The upright training distributes the bearing wood more evenly upon the vine and is thought, therefore, to insure more uniform fruit, it keeps the top near the root, which is sometimes thought to be an advantage, and it is better suited to the stature of the small-growing varieties. There is, perhaps, a greater temptation to neglect the vines in the drooping systems than in the others, because the shoots need no tying and do not, therefore, demand frequent attention; while in the upright systems the shoots soon become broken or displaced if not watched. For very large areas, or circumstances in which the best of care cannot be given the vineyard, the Kniffin or drooping systems are perhaps always to be recommended. Yet the Kniffin profits as much from diligence and skill as the other systems; but it will give better results than the others under partial neglect. The strong varieties, especially those making long and drooping canes, are well adapted to the Kniffin styles; but the smaller sorts, and those stronger sorts which, like Catawba, make an upright and stocky growth, are usually trained upon the upright systems. But the merits of both systems are so various and even so little understood, that it is impossible to recommend either one unqualifiedly. The advantages in either case are often little more than matters of personal opinion. It should be said, however, that the Kniffin or drooping systems are gaining in favor rapidly, and are evidently destined to overthrow much of the older upright training. This fact does not indicate, however, that the upright system is to be entirely superseded, but rather that. it must be confined to those varieties and conditions for which it is best adapted. The two systems will undoubtedly supplement each other. The horizontal systems are occasionally used for choice varieties, but they are little known.

Making the Trellis.—The fall or winter following the planting of the vineyard, the trellis is begun if the upright systems are used; but this operation is usually delayed a year longer in the Kniffin systems, and stakes are commonly used, or at least recommended, during the second season. In the South the trellis is made the first year. The style of trellis will depend upon the style of training, but the main features are the same for all. Strong posts of some durable timber, as cedar, locust or oak, are placed at such distance apart that two vines can be set between each two. If the vines are set nine feet apart, the posts may be eighteen or twenty feet apart, and a vine will then stand four or five feet from each post. If the posts in the row are eighteen feet apart and the rows eight feet apart, about 330 posts will be required to the acre. Except in very hard and stony lands, the posts are driven with a heavy maul, although many people prefer to set the end posts in holes, thinking that they endure the strain better. In all loose soils, however, posts can be made as firm by driving as by setting with a spade. All posts should be as firm as possible, in order to hold up the heavy loads of vines and fruit. In setting posts on hillsides, it is a common practice to lean them slightly uphill, for there is always a tendency for the posts to tilt down the slope. For the Kniffin systems, especially for the strong-growing grapes, the posts must stand six or six and one-half feet high when set, but a foot less will usually be sufficient for the upright and horizontal systems. The posts should stand higher at first than is necessary for the support of the wires, for they will need to be driven down occasionally as they become loose. The end posts of each row should be well braced, as shown in several of the illustrations in this volume.

The wire ordinarily used is No. 12, except for the

The Trellis.

top wire in the Kniffin training, which is usually No. 10, as the greater part of the weight is then upon the top wire. No. 9 is sometimes used, but it is heavier than necessary. No. 14 is occasionally used for the middle and upper rows in the upright systems, but it is not strong enough. The following figures show the sizes and weights of these and similar iron and steel wires :

No.	Diameter in inches.	Weight of 100 feet.	Feet in 2,000 pounds.
9	.148	5.80 pounds.	34,483
10	.135	4.83 "	41,408
11	.120	3.82 "	52,356
12	.105	2.92 "	68,493
13	.092	2.24 "	89,286
14	.080	1.69 "	118,343
15	.072	1.37 "	145,985
16	.063	1.05 "	190,476

The plain annealed iron wire costs about 3 cents per pound, and the galvanized—which is less used for vineyards—3½ cents. Of No. 12 wire, about 160 pounds is required per acre for a single run on rows eight feet apart, and about 500 pounds for three runs. The cost of No. 12 wire per acre, for three runs, therefore, is about $15.

The wire is secured to the intermediate posts by staples driven in firmly so that the wire will not pull through readily of its own weight, but still loosely enough to allow of the tightening of the wires. In other words, the head of the staple should not quite touch the wire. Grape staples are of three lengths, about an inch, inch and a quarter, and an inch and a half respectively. The shortest length is little used. The medium length is used for hard-wood posts and the longest for soft posts, like chestnut and cedar. These staples cost five cents per pound usually, and a pound of the medium length contains from 90 to 100 of the No. 10 wire size. An acre, for three wires, will therefore require, for this size, about nine or ten pounds of staples. In windy regions, the wires should be placed upon the windward side of the posts

There are various devices for securing the wire to the end posts, but the commonest method is to wind them about the post once and secure them with a staple, or twist the end of the wire back upon itself, forming a loop. The wires should be drawn taut to prevent sagging with the weight of fruit and leaves. In order to allow for the contraction of the wires in winter, some growers loosen the wires after harvest and others provide some device which will relieve the strain. The Yeoman's Patent Grape-Vine Trellis is a simple and effective lever-contrivance attached to each wire, and which is operated to loosen the wires in fall and to tighten them in spring. The end post is sometimes provided upon the back with a square-headed pin which works tightly in an inch and a half augur hole and about which the end of the wire is wound. A square-headed iron wrench operates the pin, while the tension of the wire around the side of the post keeps the pin from slipping. This device is not durable, however. An ingenious man can easily contrive some device for relieving the tension, if he should think it necessary. As a matter of practice, however, the wires soon stretch and sag enough with the burden of fruit and vines to take up the winter contraction, and most growers do not release the wires in fall. It will be found necessary, in fact, to tighten the wires and to straighten up the posts from year to year, as they become loose. It is always a profitable labor to tamp the ground firmly about all the posts every spring. The wires should always be kept tight during the growing season to prevent the whipping of the vines by wind. This is especially important in white grapes, which are discolored by the rubbing of leaves and twigs. Unless the vines are very strong it will be necessary to stretch only one wire the first winter.

Trellises are often made of slats, as shown in Fig. 18, but these are always less durable than the wire trellises and more expensive to keep in repair ; and in the older portions of the country, where timber is dear, they are also more expensive at the outset. They catch the wind, and, not being held together by continuous strands, are likely to blow down in sections. Fuller particulars concerning the styles of trellis are given in the discussions of the different systems of training.

Tying.—Probably the best material for tying the canes and shoots to the trellis is raffia. This is a bast-like material which comes in skeins and which can be bought of seedsmen and nurserymen for about 20 cents a pound. A pound will suffice to tie a quarter of an acre of upright training throughout the season. Raffia is obtained from the strippings of an oriental palm (*Raphia Ruffia*). Wooltwine is also still largely used for tying, but it is not so cheap and handy as raffia, and it usually has to be cut when the trellis is stripped at the winter pruning, while the raffia breaks with a quick pull of the vine. Some complain that the raffia is not strong enough to hold the vine during the season, but it can easily be doubled. Osier willows are much used for tying up the canes in the spring, and also for summer tying, especially in the nursery regions where the slender trimmings of the cultivated osier willows are easily procured. Wild willows are often used if they can be obtained handily. These willows are tied up in a small bundle, which is held upon the back above the hips by a cord passed about the body. The butts project under the right hand, if the person is right-handed, and the strands are pulled out as needed. The butt is first used, the tie being made with a twist and tuck, the strand is then cut off with a knife, and the twig is operated in like manner until it is used up. When wooltwine is used, the ball is often held in front of the workman by a cord which is tied about it and then passed about the waist. The ball is unwound from the inside, and it will hold its shape until the end becomes so short that it will easily drag upon the ground. Some workmen carry the ball in a bag,

Tying.

after the manner of carrying seed-corn. Raffia is not so easily carried in the field as the wool-twine or the willow, and this fact interferes with its popularity. Green rye-straw, cut directly from the field, is much used for tying the shoots in summer. Small wire, about two-thirds the size of broom-wire, is used occasionally for tying up the canes in spring, but it must be used with care or it will injure the vine. Corn-husks are also employed for this purpose when they can be secured. Bass-bark is sometimes used for tying, but in most of the grape regions it is difficult to secure, and it has no advantage over raffia.

It is very important that the canes be tied up early in spring, for the buds are easily broken after they begin to swell. These canes are tied rather firmly to the wires to hold them steady; but the growing shoots, which are tied during the summer, are fastened more loosely, to allow of the necessary increase in diameter.

CHAPTER III.

THE UPRIGHT SYSTEMS.

The upright systems are the oldest and best known of the styles of American grape training. They consist, essentially, in carrying out two horizontal canes, or sometimes arms, upon a low wire and training the shoots from them vertically upwards. These shoots are tied to the upper wires as they grow. This type was first clearly and forcibly described in detail by A. S. Fuller, in his "Grape Culturist," in 1864, and it became known as the Fuller system, although it was practiced many years previous to this time.

Horizontal Arm Spur System.—There are two types or styles of this upright system. The older type and the one described in the books, is known as the Horizontal Arm Spur training. In this method, the two horizontal branches are permanent, or, in other words, they are true arms. The canes are cut back each fall to upright spurs upon these arms, as explained on page 15 (fig. 4.) Two shoots are often allowed to grow from each of these spurs, as shown in fig. 7. These spurs become overgrown and weak after a few years, and they are renewed from new shoots which spring from near their base

7. HORIZONTAL ARM SPUR TRAINING.

or from the arm itself. Sometimes the whole arm is renewed from the head of the vine, or even from the ground.

The number of these upright canes and their distance apart upon these permanent arms depend upon the variety, the strength of the vine and soil and the fancy of the grower. From twelve to

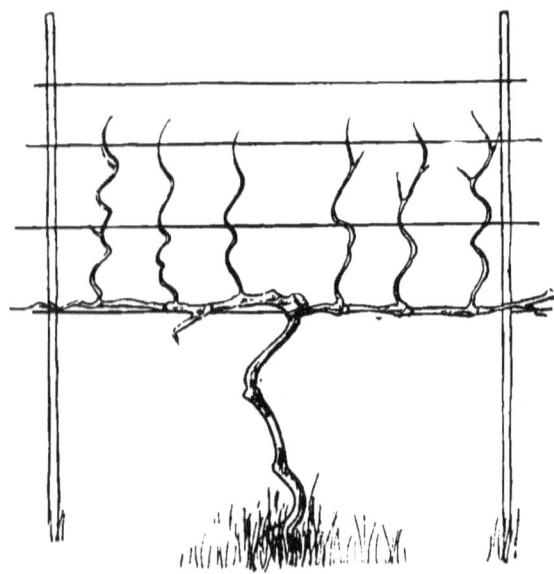

8. HORIZONTAL ARM. (Diagram.)

twenty inches apart upon the arm is the common distance. If a vine is strong enough to carry five canes and the vines are eight feet apart, then the canes are distributed at intervals of about twenty inches. Some very strong vines of vigorous varieties will carry eight canes upon the two arms together, and in this case the canes stand about a foot apart. In the fall or winter, the cane is cut away and the strongest new cane which springs from its base is left for the bearing wood of the following year. This new cane is itself headed in to the height of the trellis; that is, if the uppermost and lowermost wires are 34 inches apart—as they are in the Brocton vineyards of western New York, where this system is largely used—this new cane is shortened in to 34 inches long. Upon this length of cane there will be about seven good buds in the common varieties.

A modification of this horizontal arm system is shown in fig. 9. It is used about Forestville, in Chautauqua county, New York. The arms in this case are very short, and canes are taken out only at two or three places. The picture shows a vine in which two canes are taken from the end of each arm, making four canes for the bearing top of the vine. These canes are cut back to spurs in the fall, as explained in the above paragraph. Sometimes one or two other canes are taken out of these arms nearer the main trunk. The advantages urged for this style of training are the stronger growth which is insured by so few canes, and the small amount of old or permanent wood which is left to each vine.

The horizontal arm training is less popular than it was twenty years ago. It has serious faults, especially in the persistence of the old spurs, and

9. SHORT ARM SPUR TRAINING.

probably will eventually give place to other systems. Aside from the spur pruning, the system is much like the following, which is a modification to allow of a renewal pruning and to which the reader is referred for further details. This modification, which may be called the High Renewal, and which is one of the most serviceable of any of the styles of training, although it has never been fully described, we shall now consider.

The High Renewal, or upright training which is now very extensively employed in the lake regions of New York and elsewhere, starts the head or branches of the vine from eighteen to thirty inches from the ground. The ideal height for most varieties is probably about two feet to the first wire, although thirty inches is better than eighteen. If the vines are lower than two feet, they are liable to be injured by the plow or cultivator, the earth is dashed against the clusters by heavy rains, and if the shoots become loose they strike the ground and the grapes are soon soiled. A single trunk or arm is carried up to the required height, or if good branches happen to form lower down, two main canes are carried from this point up to the required distance to meet the lower wire, so that the trunk be-becomes Y-shaped, as seen in figs. 10, 16 and 17. In fact, vineyardists usually prefer to have this head or crotch a few inches below the lowest wire, to facilitate the spreading and placing of the canes. The trellis for the upright systems nearly always comprises three wires, although only two are sometimes used for the smaller growing varieties, and very rarely four are used for the strongest kinds, although this number is unnecessary. The lowest wire is stretched at eighteen, twenty-four or thirty inches from the ground, and the two upper ones

10. THE SECOND SEASON OF UPRIGHT TRAINING.

are placed at distances of eighteen or twenty inches apart.

The second season after planting should see the vine tied to the first wire. Fig. 10 is a photograph taken in July, 1892, of a Concord vine which was set in the spring of 1891. In the fall of 1891 the vine was cut back to three or four buds, and in the spring of 1892 two of these buds were allowed to

The Upright-System.

make canes. These two canes are now tied to the wire, which was stretched in the spring of 1892. In this case, the branches start near the surface of the ground. Sometimes only a single strong shoot grows, and in order to secure the two branches it is broken over where it passes the wire, and is usually tied to a stake to afford support. Fig. 11 shows this operation. A bud will develop at the bend or break, from which a cane can be trained in the opposite direction from the original portion, and the T-head is secured.

The close of the second season after planting, therefore, will usually find the vine with two good canes extending in opposite directions and tied to the wire. The pruning at that time will consist in cutting off the ends of these canes back to firm and strong wood, which will leave them bearing from five to eight buds. The third season, shoots will grow upright from these buds and will be tied to the second wire, which has now been supplied. Late in the third season the vine should have much the appearance of that shown in fig. 12. The third wire is usually added to the trellis at the close of the second season, at the same time that the second wire is put on; but occasionally this is delayed until the close of the third season. Some of the upright shoots may bear a few grapes this third season, but unless the vines are very strong the flower clusters should be removed; and a three-year-old vine should never be allowed to bear heavily. It must be remembered, however, that both these horizontal canes, with all their mass of

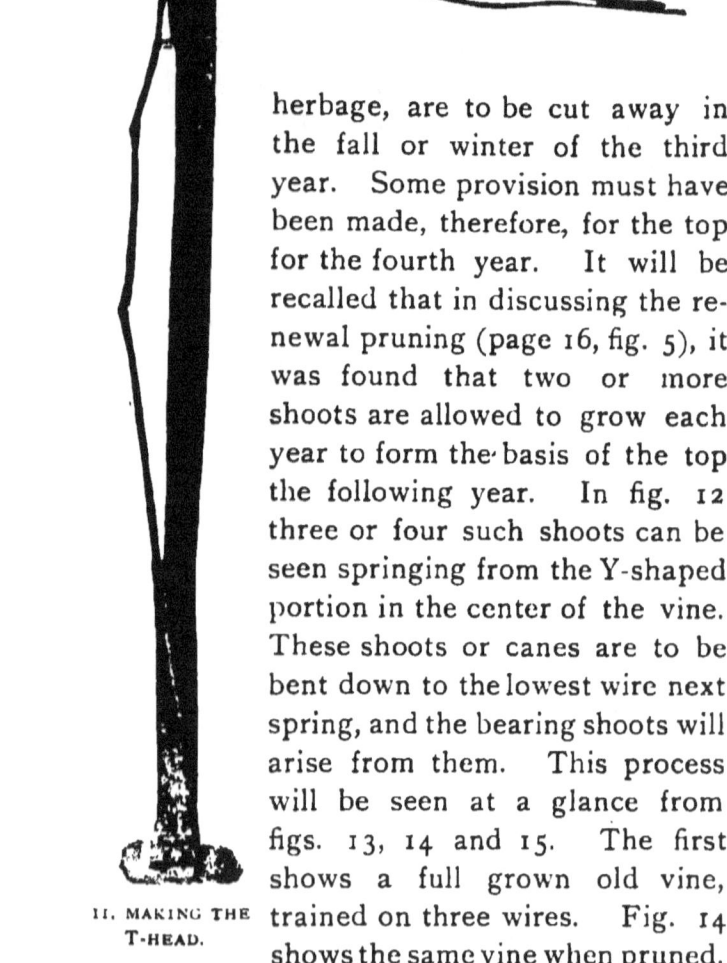

11. MAKING THE T-HEAD.

herbage, are to be cut away in the fall or winter of the third year. Some provision must have been made, therefore, for the top for the fourth year. It will be recalled that in discussing the renewal pruning (page 16, fig. 5), it was found that two or more shoots are allowed to grow each year to form the basis of the top the following year. In fig. 12 three or four such shoots can be seen springing from the Y-shaped portion in the center of the vine. These shoots or canes are to be bent down to the lowest wire next spring, and the bearing shoots will arise from them. This process will be seen at a glance from figs. 13, 14 and 15. The first shows a full grown old vine, trained on three wires. Fig. 14 shows the same vine when pruned. Two long canes, with six or eight buds each, are

12. THE THIRD SEASON OF HIGH RENEWAL.—CONCORD.

13. HIGH RENEWAL, BEFORE PRUNING.—CATAWBA.

left to form the top of the following year. The two stubs from which the renewal canes are to grow for the second year's top are seen in the center. In the fall of the next year, therefore, these two outside canes will be cut away to the base of

14. HIGH RENEWAL, PRUNED.

these renewal stubs; and the renewal canes, in the meantime, will have made a year's growth. These renewal stubs in this picture are really spurs, as will be seen; that is, they contain two ages of wood. It is the purpose, however, to remove these stubs or spurs every two or three years at most,

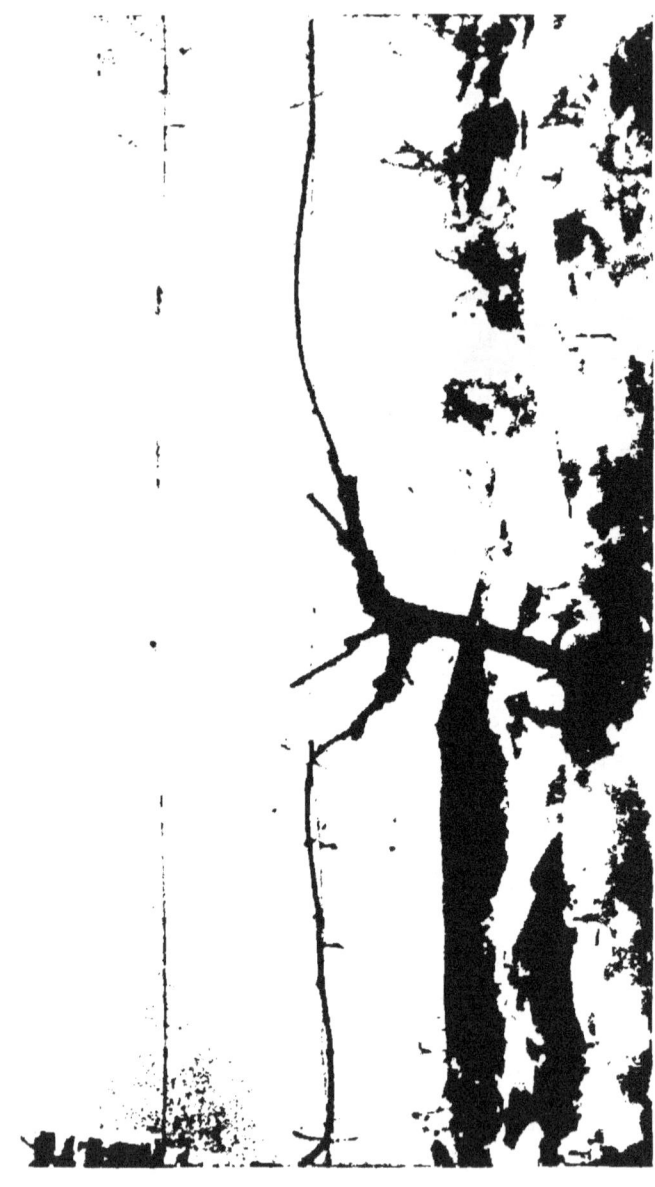
15. HIGH RENEWAL, PRUNED AND TIED.

The Upright System. 47

and to bring new canes directly from the old wood or head. If possible, the renewal cane is brought from a new place on the old wood every year in order to avoid a spur. Such was the case in the vine shown in fig. 5, page 19. Fig. 15 shows the same vine tied down to the lowest wire. Two ties have been made upon each cane. Fig. 16 shows a vine in which four canes have been left to form the top for the following year. The stubs for the renewals can be seen in the Y. It is customary to leave more than two canes, occasionally, in strong-growing varieties like Concord. Sometimes four and occasionally six are left. If four canes are left, two may be tied together in each direction upon the bottom wire. If six are used, the two extra ones should be tied along the second wire, parallel with the lowest ones. These extra canes are sometimes tied obliquely across the trellis, but this practice should be discouraged, for the usual tendency of the vine is to make its greatest growth at the top, and the lower buds may fail to bear.

16. HIGH RENEWAL WITH FOUR CANES.

The ideal length of the two canes varies with different varieties and the distance apart at which

17. HIGH RENEWAL COMPLETE.—CONCORD.

The Upright System.

the vines are set. Very strong kinds, like Concord and Niagara, can carry ten or twelve buds on each cane, especially if the vines are set more than eight feet apart. Fig. 17 shows half of a Concord vine in which about ten buds were left on each cane. These strong sorts can often carry forty or fifty buds to the vine to advantage, but when this number is left the canes should be four, as explained in the last paragraph. In Delaware and other weak-growing varieties, twenty or twenty-five buds to the vine should be the maximum and only two canes should be left. In short-jointed varieties, the canes are usually cut to the desired length—four to six feet —even if too great a number of buds is left, but the shoots which spring from these extra buds are broken out soon after they start. A Delaware vine which has made an unusually short or weak growth will require fewer buds to be left for next year's top than a neighboring vine of the same variety which has made a strong growth. The Catawba, which is a short but very stiff grower, is usually cut back to six or eight buds, as seen in figs. 13, 14 and 15. The grower soon learns to adjust the pruning to the character of the vine without effort. He has in his mind a certain ideal crop of grapes, perhaps about so many bunches, and he leaves enough buds to produce this amount, allowing, perhaps, ten per cent. of the buds for accidents and barren shoots. He knows, too, that the canes should always be cut back to firm, well-ripened wood. It should be said that mere size of cane does not indicate its value as a fruit-bearing branch. Hard, smooth wood of medium size usually gives better results than the very large and softer canes which are sometimes produced on soils rich in nitrogenous manures. This large and overgrown wood is known as a "bull cane." A cane does not attain its full growth the first year, but will increase in diameter during the second season. The tying therefore, should, be sufficiently loose or elastic to allow of growth, although it should be firm enough to hold the cane constantly in place. The cane should not be hung from the wire, but tied close to it, provision being made for the swelling of the wood to twice its diameter.

The shoots are tied to the second wire soon after they pass it, or have attained firmness enough to allow of tying, and the same shoots are tied again to the top wire. All the shoots do not grow with equal rapidity, and the vineyard must be gone over more than twice if the shoots are kept properly tied. Perhaps four times over the vineyard will be all that is necessary for careful summer tying. Many vineyardists tie only once or twice, but this neglect should be discouraged. This tying is mostly done with green rye straw or raffia. A piece of straw about ten inches long is used for each tie, it usually being wrapped but once about the shoot. The knot is made with a twist and tuck. If raffia is used, a common string-knot is made. When the

18. A SLAT TRELLIS, WITH UPRIGHT TRAINING.

shoots reach the top of the trellis, they are usually allowed to take care of themselves. The Catawba shoots stand nearly erect above the top wire and ordinarily need no attention. The long-growing varieties will be likely to drag the shoots upon the ground before the close of the season. If these tips interfere with the cultivation, they may be clipped off with a sickle or corn-cutter, although this practice should be delayed as long as possible to prevent the growth of laterals (see page 21). It is probably better to avoid cutting entirely. Some growers wind or tie the longest shoots upon the top wire, as seen in fig. 17. It is probably best, as a rule, to allow the shoots to hang over naturally, and to clip them only when they seriously interfere with the work of the hoe and cultivator. The treatment for slat trellises, as shown in fig. 18, is the same as on wire trellises, except that longer strings must be used in tying.

It is apparent that nearly or quite all the fruit in the High Renewal is borne between the first and second wires, at the bottom of the trellis. If the lower wire is twenty-four or thirty inches high, this fruit will hang at the most convenient height for picking. The fruit trays are set upon the ground, and both hands are free. The fruit is also protected from the hot suns and from frost; and if the shoots are properly tied, the clusters are not shaken roughly by the wind. It is, of course, desirable that all the clusters should be fully exposed to light

The Upright System.

and air, and all superfluous shoots should, therefore, be pulled off, as already explained (page 21). In rare cases it may also be necessary, for this purpose, to prune the canes which droop over from the top of the trellis.

After a few years, the old top or head of the vine becomes more or less weak and it should be renewed from the root. The thrifty vineyardist anticipates this circumstance, and now and then allows a thrifty shoot which may spring from the ground to remain. This shoot is treated very much like a young vine, and the head is formed during the second year (page 16, bottom). If it should make a strong growth during the first year and develop stout laterals, it may be cut back only to the lowest wire the first fall; but in other cases, it should be cut back to two or three buds, from one of which a strong and permanent shoot is taken the second year. When this new top comes into bearing, the old trunk is cut off at the surface of the ground, or below if possible. A top will retain its vigor for six or eight years under ordinary treatment, and sometimes much longer. These tops are renewed from time to time as occasion permits or demands, and any vineyard which has been bearing a number of years will nearly always have a few vines in process of renewal. The reader should not receive the impression, however, that the life or vitality of a vine is necessarily limited. Vines often continue to bear for twenty years or more without renewal; but the head after a time comes to be large and rough and crooked, and often weakened by scars, and better results are likely to be obtained if a new, clean vine takes its place.

The High Renewal is extensively used in the lake region of Western New York, for all varieties. It is particularly well adapted to Delaware, Catawba, and other weak or short varieties. When systematically pursued, it gives fruit of the highest excellence. This High Renewal training, like all the low upright systems, allows the vines to be laid down easily in winter, which is an important consideration in many parts of Canada and in the colder northern states.

Fan Training.—A system much used a few years ago and still sometimes seen, is one which renews back nearly to the ground each year, and carries the fruiting canes up in a fan-shaped manner. This system has the advantages of dispensing with much of the old wood, or trunk, and facilitating laying down the vine in winter in cold climates. On the other hand, it has the disadvantages of bearing the fruit too low—unless the lower clusters are removed—and making a vine of inconvenient shape for tying. It is little used at present. Fig. 19 shows a vine pruned for fan-training, although it is by no means an ideal vine. This vine has not been properly renewed, but bears long, crooked spurs, from which the canes spring. One of these spurs will be seen to extend beyond the lower wire. The spurs should be kept very short, and they should be entirely removed every two or three years, as explained in the above discussion of the High Renewal training.

The shoots are allowed to take their natural course, being tied to any wire near which they chance to grow, finally lopping over the top wire. Sometimes the canes are bent down and tied horizontally to the wires, and this is probably the better practice. Two canes may be tied in each direction on the lower wire, or the two inner canes may be tied down to the second wire. In either case, the vine is essentially like the High Renewal, except that the trunk is shorter.

19. FAN TRAINING, AFTER PRUNING.

CHAPTER IV.

THE DROOPING SYSTEMS.

In 1845 William T. Cornell planted a vineyard in the Hudson River Valley. A neighbor, William Kniffin, was a stone mason with a few acres of land to which he devoted his attention during the leisure seasons of his trade. Cornell induced Kniffin to plant a few grapes. He planted the Isabella, and succeeding beyond his expectations, the plantation was increased into a respectable vineyard and Kniffin came to be regarded as a local authority upon grape culture. Those were the pioneer days in commercial grape growing in North America, and there were no undisputed maxims of cultivation and training. If any system of close training and pruning was employed, it was probably the old horizontal arm spur system, or something like it. One day a large limb broke from an apple-tree and fell upon a grape-vine, tearing off some of the canes and crushing the vine into a singular shape. The vine was thought to be ruined, but it was left until the fruit could be gathered. But as the fruit matured, its large size and handsome appearance attracted attention. It was the best fruit in the vineyard! Mr. Kniffin was an observant man, and he inquired into the cause of the excellent fruit. He noticed that the vine had been pruned and that the best canes stood out horizontally. From this suggestion he developed the four-cane system of training which now bears his name. A year or two later, in 1854, the system had attracted the attention of those of his neighbors who cultivated grapes, and

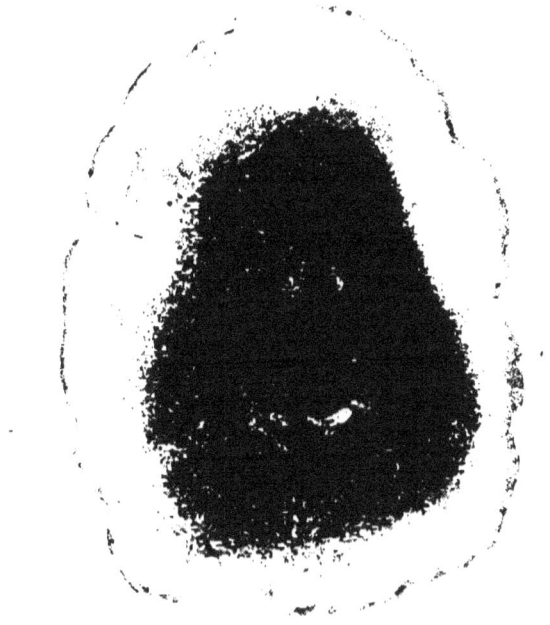

23. WILLIAM KNIFFIN.

thereafter it spread throughout the Hudson valley, where it is to-day, with various modifications, the chief method of grape training. Its merits have become known beyond its original valley, and it is now spreading more rapidly than any other system. The ground upon which the old Isabellas grew is now occupied by Concords, which are as vigorous and productive as those grown upon newer soils. William Kniffin died at his home in Clintondale, Ulster county, New York, June 13, 1876, at fifty-seven years of age. The portrait is from a photograph which was taken two or three years before his death.

The True or Four-Cane Kniffin System.—Figure 21 shows the true Kniffin system, very nearly as practiced by its originator. A single stem or trunk is carried directly to the top wire, and two canes are taken out from side spurs at each wire. Mr. Kniffin believed in short canes, and cut them back to about six buds on both wires. But most growers now prefer to leave the upper canes longer than the lower ones, as seen in illustration. The bearing shoots are allowed to hang at will, os that no summer tying is necessary; this is the distinguishing mark of the various Kniffin systems. The main trunk is tied to each wire, and the canes are tied to the wires in spring. This system possesses the great advantage, therefore, of requiring little labor during the busy days of the growing season ; and the vines are easily cultivated, and if the rows are nine or ten feet apart, currants or other bush-fruits can be grown between. The system is especially adapted to the strong varieties of grapes. For further comparisons of the merits of different systems of training, the reader should consult Chapter II.

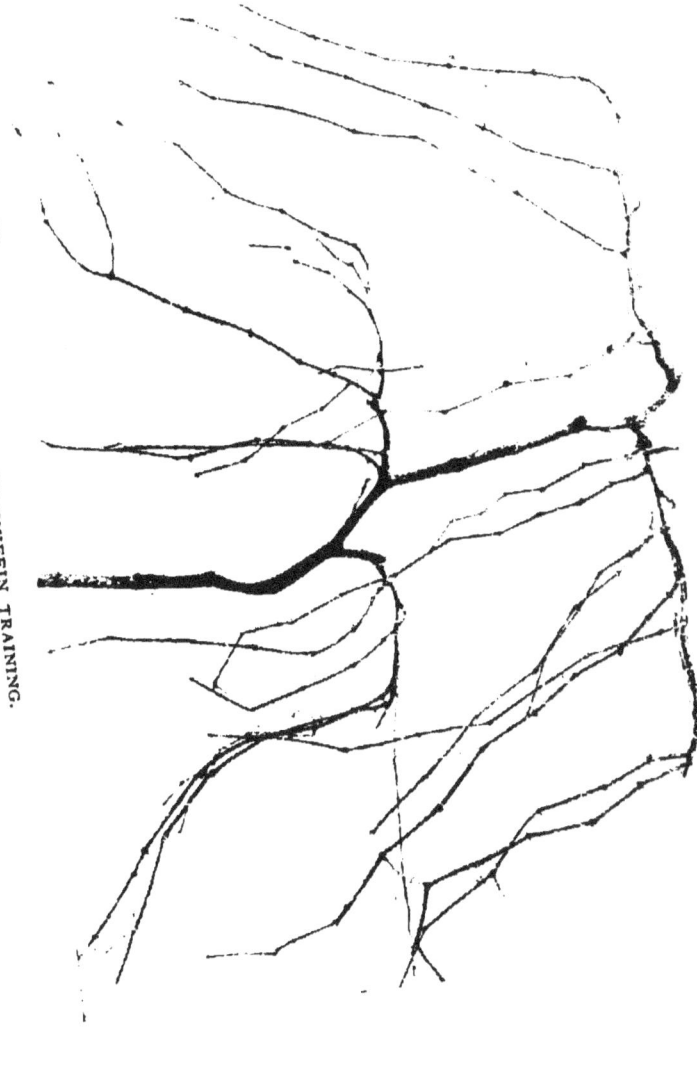

21. THE TRUE KNIFFIN TRAINING.

The pruning of the Kniffin vine consists in cutting off all the wood save a single cane from each spur. Fig. 22 illustrates the process. This is the same vine which is shown with the full amount of wood on in fig. 21. The drooping shoots shown in that illustration bore the grapes of 1892; and now, in the winter of 1892-93, they are all to be cut away, with the horizontal old canes from which they grew, save only the four canes which hang nearest the main trunk. Fig. 22 shows the vine after it had been pruned. It is not obligatory that the canes which are left after the pruning should be those nearest the trunk, for it may happen that these may be weak; but, other things being equal, these canes are preferable because their

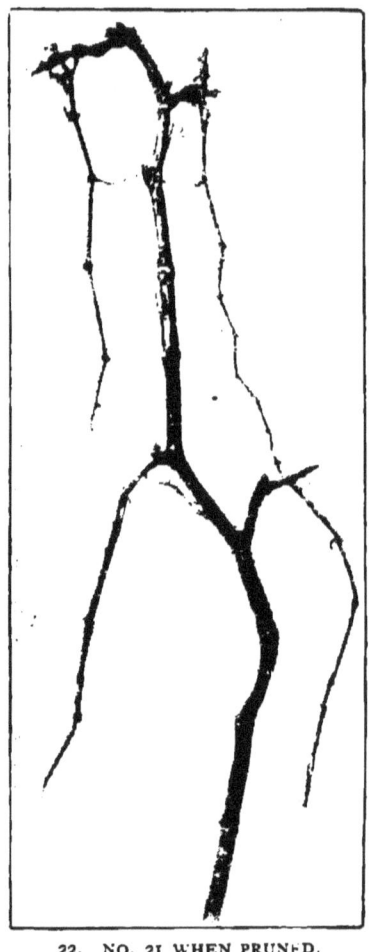

22. NO. 21 WHEN PRUNED.

selection keeps the old spurs short. The careful grower will take pains to remove the weak shoots which start from this point, in order that a strong cane may be obtained. It is desirable that these side spurs be removed entirely every three or four years, a new cane being brought out again from the main body or trunk. There is little expectation, however, that there shall be such a complete renewal pruning as that practiced in the High Renewal, which we discussed in the last chapter.

It will be seen that the drooping canes in fig. 22 are shorter than they were originally, as shown in fig. 21. They have been cut back. The length at which these canes shall be left is a moot point. Much depends upon the variety, the distance between the wires, the strength of the soil, and other factors. Nearly all growers now agree that the upper canes should be longer than the lower ones, although equal canes are still used in some places. In strong varieties, like Worden, each of the upper canes may bear ten buds and each of the lower ones five. This gives thirty buds to the vine. Some growers prefer to leave twelve buds above and only four below.

These four pruned canes are generally allowed to hang during winter, but are tied onto the wires before the buds swell in spring. They are stretched out horizontally and secured to the wire by one or two ties upon each cane. The shoots which spring from these horizontal canes stand upright or oblique at first but they soon fall over with the weight of foliage and fruit. If they touch the ground, the ends may be clipped off with a sickle, corn-cutter or scythe, although this is not always done, and is not necessary unless the canes interfere with cultivation. There is no summer-pinching nor pruning, although the superfluous shoots should be broken out, as in other systems. (See page 23).

Only two wires are used in the true Kniffin trellis. The end posts are usually set in holes, rather than driven, to render them solid, and they should always be well braced. The intermediate posts are driven, and they usually stand between every alternate vine, or twenty feet apart if the vines are ten feet apart—which is a common distance for the most vigorous varieties. For the strong-growing varieties, the top wire is placed from five and one-half to six feet above the ground. Five feet nine inches is a popular height. The posts will heave sufficiently to bring the height to six feet, although it is best to "tap" the posts every spring with a maul in order to drive them back and make them firm. The lower wire is usually placed at three and one-half feet. Delawares, if trained Kniffin, should not stand above five feet four inches, or at most five feet six inches. Strong vines on good soil are often put onto the trellis the second year, although it is a commoner practice, perhaps, to stake them the second season, as already explained (page 27), and put them on the wires the third season. The year following the tying to the trellis,, the vine should bear a partial crop. The vine is usually carried directly to the top wire the first season of training, although it is the practice of some growers, especially outside the Hudson valley, to stop the trunk at the lower wire the first year of permanent training, and to carry it to the top wire the following year.

Yields from good Kniffin vines will average fully as high and perhaps higher than from other species of training. W. D. Barns, of Orange county, New York, has had an annual average of twenty-six pounds of Concords to the vine for nine years, 1,550 vines being considered in the calculation. While the Delaware is not so well suited to the Kniffin system as stronger varieties, it can nevertheless be trained in this manner with success, as the following average yields obtained by Mr. Barns from 200 vines set in 1881 will show:

    1886 . . . . 8½ pounds to the vine.
    1887 . .   . . .11¾   "   "   "   "
    1888 8   "   "   "   "
    1889 . . .   . 9½   "   "   "   "
    1890 7   "   "   "   "
    1891 16   "   "   "   "
    1892 13   "   '   "   "

Modifications of the Four-Cane Kniffin.—Various modifications of this original four-cane Kniffin are in use. The Kniffin idea is often carelessly applied to a rack trellis. In such cases, several canes were allowed to grow where only two should have been left. Fig. 23 is a common but poor style

23. A POOR TYPE OF KNIFFIN.

of Kniffin used in some of the large new vineyards of western New York. It differs from the type in the training of the young wood. These shoots, instead of being allowed to hang at will, are carried out horizontally and either tied to the wire or

24. THE Y-TRUNK KNIFFIN.

twisted around it. The advantage urged for this modification is the little injury done by wind, but, as a matter of practice, it affords less protection than the true drooping Kniffin, for in the latter the shoots from the upper cane soon cling to the lower wire, and the shoots from both tiers of canes protect each other below the lower wire. There are three serious disadvantages to this holding up of the shoots,—it makes unnecessary labor, the canes are likely to make wood or "bull canes" (see page 50) at the expense of fruit, and the fruit is bunched together on the vines.

Another common modification of the four-cane Kniffin is that shown in fig. 24, in which a crotch or Y is made in the trunk. This crotch is used in the belief that the necessary sap supply is thereby more readily deflected into the lower arms than by the system of side spurring on a straight or continuous trunk. This is probably a fallacy, and may have arisen from the attempt to grow as heavy canes on the lower wire as on the upper one. Nevertheless, this modification is in common use in western New York and elsewhere.

If it is desired to leave an equal number of buds on both wires, the Double Kniffin will probably be found most satisfactory. Two distinct trunks are brought from the root, each supplying a single wire only. The trunks are tied together to hold them in place. This system, under the name of Improved Kniffin, is just coming into notice in restricted portions of the Hudson valley.

The Two-Cane Kniffin, or Umbrella System.—Inasmuch as the greater part of the fruit in the Four-Cane Kniffin is born upon the upper wire, the question arises if it would not be better to dispense with the lower canes and cut the upper ones longer. This is now done to a considerable extent, especially in the Hudson valley. Fig. 25 explains the operation. This shows a pruned vine. The trunk is tied to the lower wire to steady it, and two canes, each bearing from nine to fifteen buds, are left upon the upper wire. These canes are tied to the upper wire and they are then bent down, hoop-like, to the

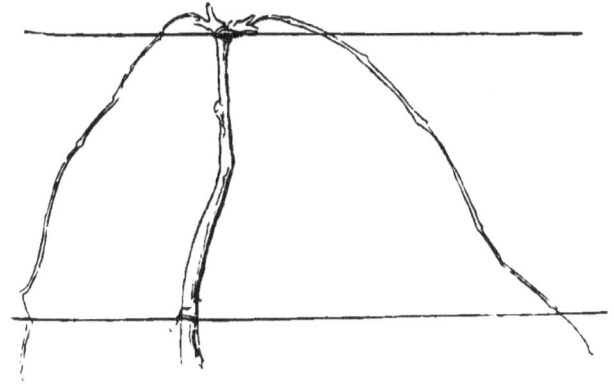

25. UMBRELLA TRAINING.

lower wire, where the ends are tied. In some instances, the lower wire is dispensed with, but this is not advisable. This wire holds the vine in place against the winds and prevents the too violent whipping of the hanging shoots. During the growing season, renewal canes are taken from the spurs in exactly the same manner as in the ordinary Kniffin. This species of training reduces the amount of leaf-surface to a minimum, and every precaution must be taken to insure a healthy leaf-growth. This

26. A POOR UMBRELLA SYSTEM.

system of training will probably not allow of the successful girdling of the vine for the purpose of hastening the maturity and augmenting the size of the fruit. Yet heavy crops can be obtained from it, if liberal fertilizing and good cultivation are employed, and the fruit is nearly always first-class. A Concord vine trained in this manner produced in 1892 eighty clusters of first quality grapes, weighing forty pounds.

Another type of Umbrella training is shown in fig. 26, before pruning. Here five main canes were allowed to grow, instead of two. Except in very strong vines, this top is too heavy, and it is probably never so good as the other (fig. 25), if the highest results are desired; but for the grower who does not care to insure high cultivation it is probably a safer system than the other.

The Low, or One-Wire Kniffin.—A modification of this Umbrella system is sometimes used, in which the trellis is only three or four feet high and comprises but a single wire. A cane of ten or a dozen buds is tied out in each direction, and the shoots are allowed to hang in essentially the same manner as in the True or High Kniffin system. The advantages urged for this system are the protection of the grapes from wind, the large size of the fruit due to the small amount of bearing wood, the ease of laying down the vines, the readiness with which the top can be renewed from the root as occasion demands, and the cheapness of the trellis.

The Six-Cane Kniffin.—There are many old vineyards in eastern New York which are trained upon a six-cane or three-wire system. The general prunning and management of these vines do not differ from that of the common Kniffin. Very strong varieties which can carry an abundance of wood, may be profitable upon this style of training, but it cannot be recommended. A Concord vineyard over thirty years old, comprising 295 vines, trained in this fashion, is still thrifty and productive. Twice it has produced crops of six tons.

Eight-Cane Kniffin.—Eight and even ten canes are sometimes left upon a single trunk, and are trained out horizontally or somewhat obliquely, as

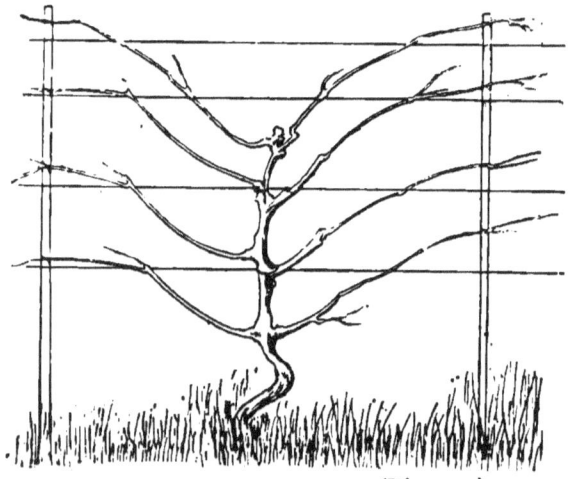

27. EIGHT-CANE KNIFFIN. (Diagram.)

shown in the accompanying diagram (fig. 27). Unless these canes are cut back to four or five buds each, the vine carries too much wood and fruit.

28. OVERHEAD KNIFFIN.

This system allows of close planting, but the trellis is too expensive. The trunk soon becomes overgrown with spurs, and it is likely to become prematurely weak. This style is very rarely used.

Overhead, or Arbor Kniffin.—A curious modification of the Kniffin is employed somewhat on the Hudson, particularly by Sands Haviland at Marlboro'. The vines are carried up on a kind of over-

29. OVERHEAD KNIFFIN.

head arbor, as shown in figs. 28, 29 and 30. The trellis is six feet above the ground, and is composed of three horizontal wires lying in the same plane. The central wire runs from post to post, and one upon either side is attached to the end of a three-foot cross-bar, as represented in fig. 28. The rows are nine feet apart, and the vines and posts twelve feet apart in the row. Contiguous rows are braced by a connecting-pole, as in fig. 29. The trunk of

30. OVERHEAD KNIFFIN, BEFORE PRUNING.

the vine ends in a T-shaped head, which is well displayed in the vine at the extreme right in the foreground in fig. 30. From this T-head, five canes are carried out from spurs. It was formerly the practice to carry out six canes, one in each direction upon each wire, but this was found to supply too much wood. Now two canes are carried in one direction and three in the other ; and the positions of these sets are alternated each year, if possible. The canes which are left after the winter pruning are tied along the wires in spring, as in the Kniffin, and the shoots hang over the wires. The chief advantage of this training is that it allows of the growing of bush-fruits between the rows, as seen in fig. 29. It is also said that the clusters hang so free that the bloom is not injured by the twigs or leaves, and the fruit is protected from sun and frost. Every post must be large and firmly set, however, adding much to the cost of the trellis. Several styles similar to this are in use, one of the best being the Crittenden system, of Michigan. In this system, the trellis is low, not exceeding four or five feet, and the vines cover a flat-topped platform two or three feet wide.

The Cross-Wire System.—Another high Kniffin training, and which is also confined to the vicinity of Marlboro', New York, is the Cross-Wire, represented in figs. 31 and 32. Small posts are set eight feet apart each way, and a single wire runs from the top of post to post—six and one-half feet from the ground—in each direction, forming a check-row system of overhead wires. The grape-vine is set at the foot of the stake, to which the trunk is tied for support. Four canes are taken from spurs on the head of the trunk, one for each of the radiating wires. These canes are cut to three and one-half or four feet in length, and the bearing shoots droop as

31. CROSS-WIRE TRAINING.

they grow. Fig. 31 shows this training as it appears some time after the leaves start in spring. Later in the season the whole vineyard becomes a great arbor, and a person standing at a distance sees an almost impenetrable mass of herbage, as in fig. 32. This system appears to have little merit, and will always remain local in application. It possesses the advantage of economy in construction of the trellis, for very slender posts are used, even at the ends of

32. CROSS-WIRE TRAINING. OUTSIDE VIEW.

the rows. The end posts are either braced by a pole or anchored by a wire taken from the top and secured to a stake or stone eight or ten feet beyond, outside the vineyard.

Renewal Kniffin.—It is an easy matter to adapt the Kniffin principle of free hanging shoots to a true renewal method of pruning. There are a few modifications in use in which the wood is annually renewed to near the ground. The trellises comprise either two or three wires, and are made in the same manner as for the upright systems, as the High Renewal. At the annual pruning only one cane is left. This comprises twelve or fifteen buds, and is tied up diagonally across the trellis, the point or end of the cane usually being bent downward somewhat, in order to check the strong growth from the uppermost parts. The shoots hang from this cane, and they may be pinched back when they reach the ground. In the meantime a strong shoot is taken out from the opposite side of the head—which usually stands a foot or less from the ground—to make the bearing wood of the next year; and this new cane will be tied in an opposite direction on the trellis from the present bearing cane, and the next renewal shoot will be taken from the other side of the head, or the side from which the present bearing wood sprung; so that the bearing top of the vine is alternated in either direction upon the trellis. This system, and similar ones, allow of laying down the vines easily in winter, and insure excellent fruit because the amount of bearing wood is small; but the crop is not large enough to satisfy most demands.

The Munson System.—An unique system of training, upon the Kniffin principle, has been devised by T. V. Munson, of Denison, Texas, a well-known authority upon grapes. Two posts are set in the same hole, their tops diverging. A wire is stretched along the top of these posts and a third one is hung between them on cross-wires. The trunk of the vine, or its head, is secured to this middle lower wire and the shoots lop over the side wires. The growth, therefore, makes a V-shaped or trough-like mass of herbage. Fig. 33 is an end view of this trellis, showing the short wire connecting the posts and which also holds the middle trellis-wire at the point of the V. Fig. 34 is a side view of the trellis. The bearing canes, two or four, in number, which are left after the annual pruning, are tied along this middle wire. The main trunk forks just under the middle wire, as seen at the left in fig. 34. A head is formed at this place not unlike

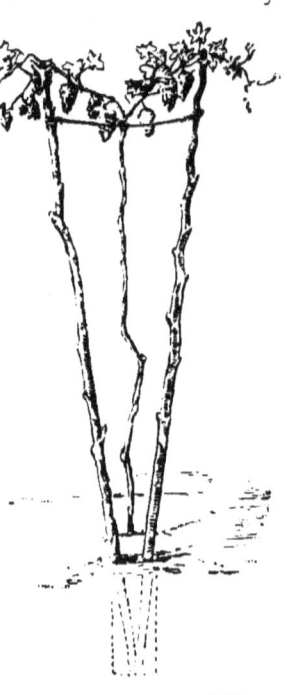

33. MUNSON TRAINING.
END VIEW.

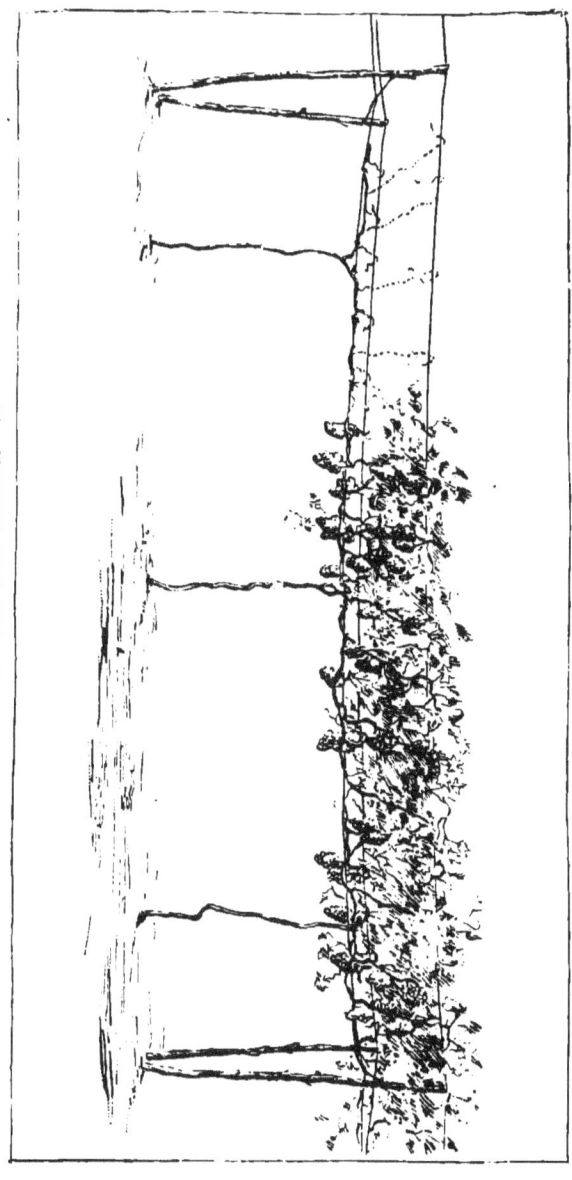

34. MUNSON TRAINING. SIDE VIEW.

like that which characterizes the High Renewal, for this system also employs renewal pruning. The trellis stands six feet high. The shoots stand upright at first, but soon fall down and are supported by the side wires. The following account of this system of training is written for this occasion by Mr. Munson:

"After the vines have flowered, the bearing laterals have their tips pinched off, and that is all the summer pruning the vine gets, except to rub off all eyes that start on the body below the crotch. Two to four shoots, according to strength of vine, are started from the forks or crotch and allowed to bear no fruit, but are trained along over the lower central wire for renewal canes. When pruning time arrives, the entire bearing cane of the present year, with all its laterals, is cut away at a point near where the young renewal shoots have started, and these shoots are shortened back, according to strength of vine; some, such as Herbemont, being able at four years to fill four shoots six or eight feet long with fine fruit, while Delaware could not well carry over three or four feet each way of one shoot only. The different varieties are set at various distances apart, according as they are strong or weak growers.

"Thus the trellis and system of pruning are reduced to the simplest form. A few cuts to each vine cover all the pruning, and a few ties complete the task. A novice can soon learn to do the work well. The trunk or main stem is secured to the middle lower wire, along which all bearing canes

The Drooping System.

are tied after pruning, and from which the young laterals which produce the crop are to spring. These laterals strike the two outer wires, soon clinging to them with their tendrils, and are safe from destruction, while the fruit is thrown in the best possible position for spraying and gathering, and is still shaded with the canopy of leaves. I have now used this trellis five years upon ten acres of mixed vines, and I am more pleased with it every year.

"The following advantages ars secured by this system:

"1. The natural habit of the vine is maintained, which is a canopy to shade the roots and body of vine and the fruit, without smothering.

"2. New wood, formed by sap which has never passed through bearing wood, is secured for the next crop—a very important matter.

"3. Simplicity and convenience of trellis, allowing free passage in any direction through the vineyard; circulation of air without danger of breaking tender shoots; ease of pruning, spraying, cultivation, harvesting.

"4. Perfect control in pruning of amount of crop to suit capacity of vine.

"5. Long canes for bearing, which agrees exactly with the nature of nearly all our American species far better than short spurs.

"6. Ease of laying down in winter. The vine

being pruned and not tied, standing away from posts, can be bent down to one side between the rows, and earth thrown upon it, and can be quickly raised and tied in position.

"7. Cheapness of construction and ease of removing trellis material and using it again.

"8. Durability of both trellis and vineyard."

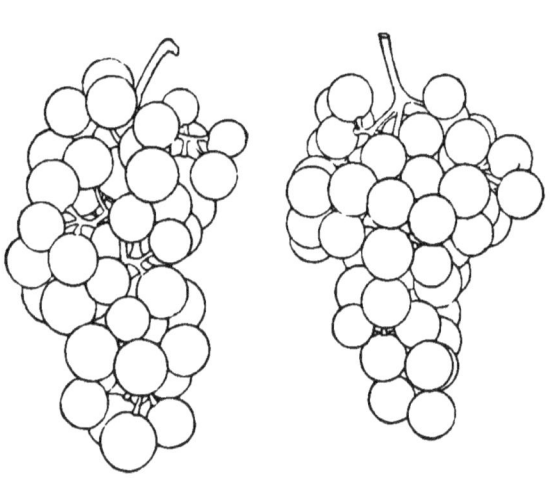

CHAPTER V.
MISCELLANEOUS SYSTEMS.

Horizontal Training.—There are very few types of horizontal shoot training now in use. The best is probably that shown in fig. 35. This particular

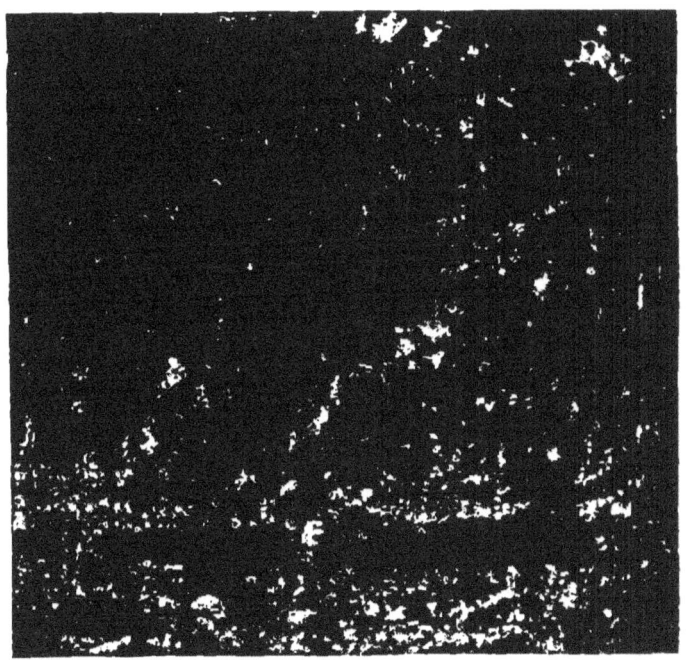

35. HORIZONTAL TRAINING.

vine is a Delaware, to which this training is well adapted. It will be noticed that this picture represents the end of a trellis, and the diagonal stick seen near the ground is a brace for the end post. Two wires run from post to post, one about two and one-half feet above the ground and the other five and one-half feet high. The posts are set at the ordinary distance of 16 or 18 feet apart. The vines are set six or eight feet apart, if Delawares. A strong stake is driven in the ground behind each vine, standing as high as the top of the trellis when set. The permanent trunk or head of the vine stands about a foot high. The vine is renewed back to the top of this trunk every year. One cane is left at each pruning, which, when tied up to the stake, is as high as the trellis. From this perpendicular cane, the bearing shoots are carried out horizontally. About six of these shoots are allowed to grow upon either side of the cane. As the shoots grow, they are tied to perpendicular slats which are fastened on the wires. These slats do not touch the ground. Two slats are provided upon either side, making four to a vine. They stand a foot or fifteen inches apart. The clusters hang free from the horizontal shoots. If the shoots grow too long, they are pinched in when they have passed the second slat. While these shoots are covering the trellis, another shoot is taken out from the head or trunk of the vine and, without being allowed to fruit, is tied up along the central stake. This shoot is to form the top next year, for all the present vine is to be entirely cut away at the winter's pruning. So the vine starts every spring with but a single cane.

Excellent results are obtained from the slender growing varieties by this method of training, but it is too expensive in trellis and in labor of tying to make it generally practicable. Delaware, however, thrives remarkably well when trained in this fashion.

Post Training.—There are various methods of training to posts, all of which possess two advantages—the saving of the expense of trellis and allowing of cultivation both ways. But they also have grave disadvantages, especially in the thickness of the head of foliage which harbors rot and mildew and prevents successful spraying, and hinders the fruit from coloring and ripening well. These faults are so serious that post training is now little used for the American grapes. The saving in cost of trellis is not great, for more posts are required to the acre than in the trellis systems, and they do not endure long when standing alone with the whole weight of the vines thrown upon them.

There are various methods of pruning for the stake training, but nearly all of them agree in pruning to side spurs upon a permanent upright arm which stands the full height of the vine. There may be one or two sets of these spurs. We might suppose the Kniffin vine, shown in fig. 22, to be tied to a post instead of stretched on a trel-

36. LOW POST TRAINING.

lis; in that event, the four canes would hang at will, or they might be wrapped about the post, the shoots hanging out unsupported in all directions. The post systems are essentially Kniffin in principle, for the shoots hang free. In low styles of post training, the permanent head of the vine may be only three or four feet high. This head will have a ring of spurs on it, and at the annual pruning three to five canes with from six to ten buds each are left. Fig. 36 is a view in such a post vineyard.

The main trunk is usually tied permanently to the post. The canes left after pruning are variously disposed. Sometimes they are bent upwards and tied to the post above the head of the vine, but they are oftenest either wound loosely about the post, or are allowed to hang loose. Two trunks are frequently used to each post, both coming from the ground from a common root. These are wound about the post in opposite directions, one outside the other, and if the outside one is secured at the top by a small nail driven through it, or by a cord, no other tying will be necessary. Sometimes two or three posts are set at distances of one foot or more apart, and the vines are wrapped about them, but this only arguments the size and depth of the mass of foliage. Now and then one sees a careful post training, in which but little wood is left and vigorous breaking out of shoots practiced, which gives excellent results;

but on the whole, it cannot be recommended. The European post and stake systems or modifications of them, are yet occasionally recommended for American vines, but under general conditions, especially in commercial grape growing, they rarely succeed long. One of the latest recommendations of any of these types is that of the single pole system of the Upper Rhine Valley, by A. F. Hofer, of Iowa, in a little treatise published in 1878.

Arbors.—Arbors and bowers are usually formed with little reference to pruning and training. The first object is to secure shade and seclusion, and these are conditions which may seriously interfere with the production of fine grapes. As a rule, too much wood must be allowed to grow, and the soil about arbors is rarely ever cultivated. Still, fair results in fruit can be obtained if the operator makes a diligent use of the pruning shears. It is usually best to carry one main or permanent trunk up to the top or center of the arbor. Along this trunk at intervals of two feet or less, spurs may be left to which the wood is renewed each year. If the vines stand six feet apart about the arbor—which is a satisfactory distance—one cane three feet long may be left on each spur when the pruning is done. The shoots which spring from these canes will soon cover up the intermediate spaces. At the close of the season, this entire cane with its laterals is cut away at the spur, and another three-foot cane—which grew during the season—is left in its place. This pruning is essentially that of the Kniffin vine in fig. 22. Imagine this vine, with as many joints or tiers as necessary, laid upon the arbor. The canes are tied out horizontally to the slats instead of being tied on wires. This same system—running up a long trunk and cutting in to side spurs—will apply equally well to tall walls and fences which it is desired to cover. Undoubtedly a better plan, so far as yield and quality of fruit is concerned, is to renew back nearly to the root, bringing up a strong new cane, or perhaps two or three every year, and cutting the old ones off; but as the vines are desired for shade one does not care to wait until midsummer for the vines to reach and cover the top of the arbor.

Remodeling Old Vines.—Old and neglected tops can rarely be remodeled to advantage. If the vine is still vigorous, it will probably pay to grow an entirely new top by taking out a cane from the root. If the old top is cut back severely for a year or two, this new cane will make a vigorous growth, and it can be treated essentially like a new or young vine. If it is very strong and ripens up well, it can be left long enough the first fall to make the permanent trunk; but if it is rather weak and soft, it should be cut back in the fall or winter to two or three buds, from one of which the permanent trunk is to be grown the second season. Thereafter, the instructions which are given in the preceding pages for the various systems, will apply to the new vine. The old trunk should be cut away as soon as the new one is permanently tied to the wires, that is, at the close of either the first or second season of the new trunk. Care must be exercised to rub off all sprouts which spring from the old root or stump. If this stump can be cut back into the ground and covered with earth, better results may be expected. Old vines treated in this manner often make good plants, but if the vines are weak and the soil is poor, the trouble will scarcely pay for itself.

These old vines can be remodeled easily by means of grafting. Cut off the trunk five or six inches below the surface of the ground, leaving an inch or two of straight wood above the roots. Into this stub insert two cions exactly as for cleft-grafting the apple. Cions of two or three buds, of firm wood the side of a lead-pencil, should be inserted. The top bud should stand above the ground. The cleft will need no tying nor wax, although it is well to place a bit of waxed cloth or other material over the wound to keep the soil out of it. Fill the earth tightly about it. Fig. 37 shows the first year's growth from two cions of Niagara set in a Red Wyoming root. Great care must be taken in any pruning which is done this first year, or the cions may be loosened. If the young shoots are tied to a stake there will be less danger from wind and careless workmen. In the vine shown in the illustration, no pruning nor rubbing out was done,

37. A YEARLING GRAFT.

but the vine would have been in better shape for training if only one or two shoots had been allowed to grow. Such a vine as this can be carried onto the trellis next year; or it may be cut back to three or four buds, one of which is allowed to make the permanent trunk next year, like a two-year set vine.

If it is desired, however, to keep the old top, it will be best to cut back the annual growth heavily at the winter pruning. The amount of wood which shall be left must be determined by the vigor of the plant and the variety, but three or four canes of six to ten buds each may be left at suitable places. During the next season a strong shoot from the base of each cane may be allowed to grow, which shall form the wood of the following season, while all the present cane is cut away at the end of the year. So the bearing wood is renewed each year, as in the regular systems of training. Much skill and experience are often required to properly rejuvenate an old vine; and in very many cases the vine is not worth the trouble.

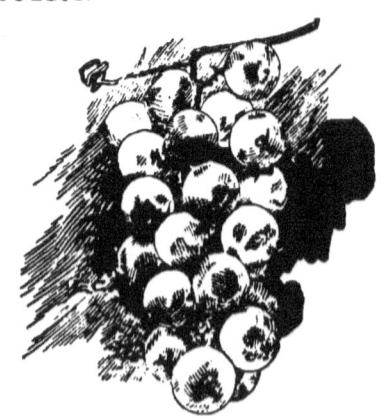

INDEX.

	Page
Adlum, quoted	10
Arbor Kniffin	72
Arbors	88
Arm, defined	13
Barns, W. D., quoted	53
Bass bark	33
Bleeding	22
Breaking-out	23
Brocton, Training at	37
Bull cane	50, 66
Cane, defined	13
Chautauqua County, Training in	37
Contraction of wires	30
Cornell, William T	56
Cornhusks, for tying	33
Crittenden training	74
Cross-wire training	74
Crotch Kniffin	66
Double Kniffin	66
Drooping systems	56
Eight-cane Kniffin	70
Fan training	54
Forestville, Training at	37
Four-cane Kniffin	58
Fuller, quoted	10, 34
Girdling	69
Grafting	90
Haviland, Sands	72
Heading-in	23
High Renewal training	39
Hofer, A. F	88
Horizontal Arm training	34
Horizontal training	83
Husks, for tying	33

Index.

	Page
Improved Kniffin	66
Kniffin systems	58
Kniffin training, Comparison of	26
Kniffin, William	56
Low Kniffin	69
Marlboro', Training at	72, 74
Modified Kniffin	63
Munson training	78
Munson, T. V.	78
Objects of pruning	24
Old vines, Remodeling of	89
One-wire Kniffin	69
Overhead Kniffin	72
Planting	20
Posts	28
Post training	85
Pruning	11
Pruning, Objects of	24
" of young vines	20
" Summer	23
" Time for	22
Raffia	32
Raphia Ruffia	32
Reasons for pruning	24
Remodeling old vines	89
Renewal, defined	18
Renewal Kniffin	77
Rubbing off	14, 23
Rye straw for tying	33
Sagging of wires	30
Setting	20
Shoot, defined	13
Six-cane Kniffin	70
Spur, defined	17
Spur training	34
Staples	29
Stopping	23
Stripping	22
Summer pruning	23
Superfluous shoots	23

Index.

	Page
Systems compared	25
T-head	41
Thomas' Fruit Culturist, quoted	10
Tightening wires	31
Trellis, Making	27
True Kniffin	58
Twine for tying	32
Two-cane Kniffin	66
Tying	31
Umbrella training	66
Upright training	34
Walls, Training on	89
Weeping	22
Willows, for tying	32
Wire, for trellis	28
" for tying	33
" weights and sizes	29
Wool-twine	32
Y-trunk Kniffin	66
Yeoman's patent trellis	30
Yields of grapes	14, 63, 69, 70
Young vines, Pruning of	20

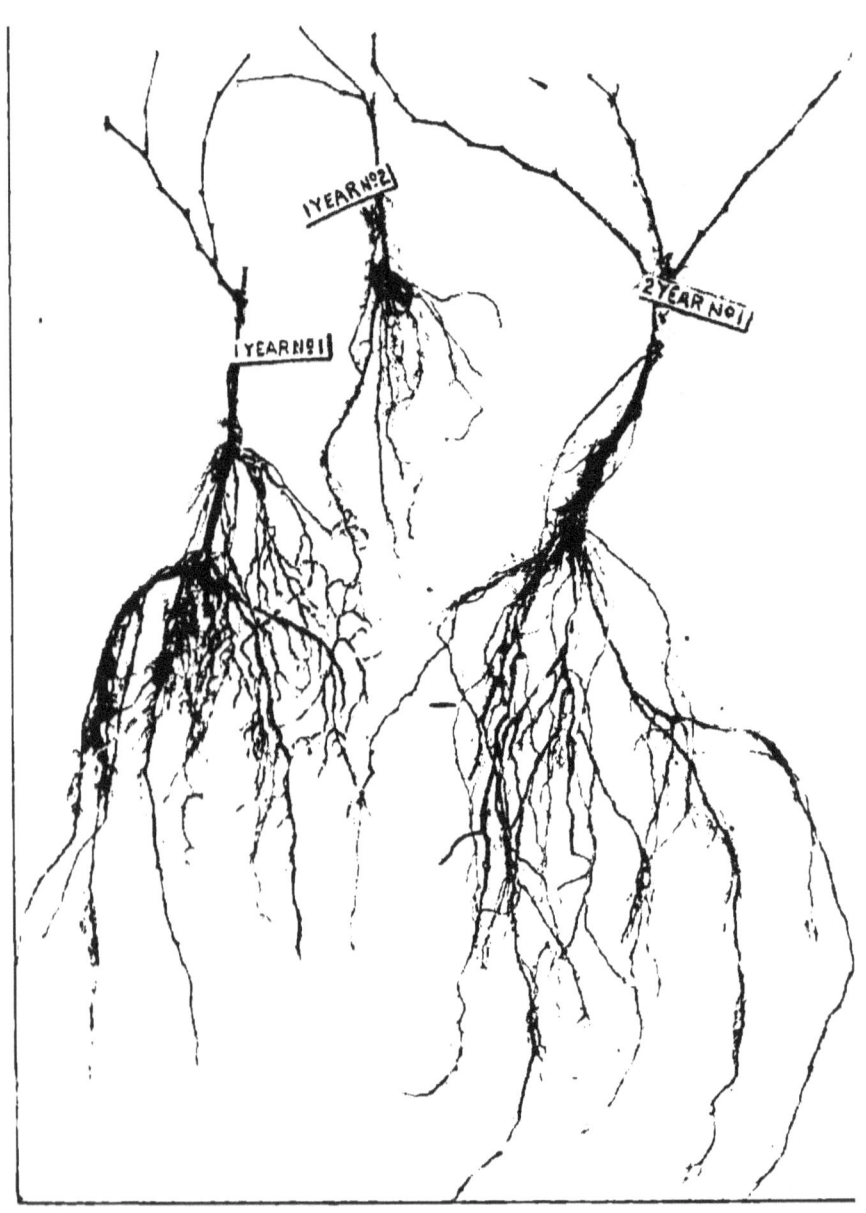

THIS ILLUSTRATION was made from a photograph of fair samples of the different g of our grape vines, reduced to one tenth their natural size.
We take great pride and confident in our ability to furnish *strong, fibrous rooted* sto(well appreciated by intelligent and experienced fruit growers.
WHOLESALE TRADEE SPECIALLY SOLICITED. CATALOGUE FR
LEWIS ROESCH, FREDONIA, N. Y.,
General Grape Vine Specialist and General Nurserymen.
When writing name this book.:

Hardy Native Grapes.

We desire to call the attention of planters to our large and complete stock of Grape Vines.

We propagate and offer for sale upwards of sixty varieties, embracing the popular old sorts as well as the new ones which seem to have merit. Our catalogue contains accurate descriptions, and classifies the different varieties according to color.

Besides the above we offer an immense collection of all kinds of Fruit and Ornamental Trees, Shrubs, Roses, Hardy Plants, etc. Our General Catalogue (160 pages), embellished with numerous engravings of the most popular Trees, Shrubs, etc., and enclosed in an illuminated cover, will be mailed free to all who have not received it.

Our Supplementary Catalogue (28 pages) of Rare and Choice Trees, Shrubs, etc., including several valuable novelties and many specialties of superior merit, will also be mailed free.

ELLWANGER & BARRY,
Mount Hope Nurseries,
ROCHESTER, N. Y.

53rd Year.

Pleasant Valley Nurseries

PEAR TREES.—Lincoln, Coreless, Bessemianka, Japan Golden Russet, Kieffer, LeConte, etc., Nut Trees in variety. Fruit Trees of all sorts. Ornamentals, Eleagnus Longipes, Japanese Wineberry Juneberry, Trifoliate Orange and other valued novelties.

FRUIT TREES! BERRY PLANTS!

STRAWBERRIES, Van Deman, E. P. Roe, and other new varieties; all the old standard sorts, Gooseberries, Raspberries, Blackberries, Currants, Asparagus Roots and Grape Vines.

J. S. COLLINS & SON, Moorestown, N. J.
Send for Catalogue.

Miscellaneous Books.
For the Farm and Household.

Any one of these valuable books will be sent, postpaid, direct, on receipt of price. Be careful to write name and post office plain, so that there may be no mistake in mailing. Address

The Rural Publishing Co., New York.

POPULAR ERRORS ABOUT PLANTS.—By A. A. CROZIER. A collection of errors and superstitions entertained by farmers, gardeners and others, together with brief scientific refutations. Highly interesting to students and intelligent readers of the new and attractive in rural literature, and of real value to practical cultivators who want to know the truth about their work.

Price, cloth, $1.

THE NURSERY BOOK.—By L. H. BAILEY. A complete handbook of Propagation and Pollination of Plants. *Profusely illustrated.* This valuable little manual has been compiled with great pains. The author has had unusual facilities for its preparation, having been aided by many experts. The book is absolutely devoid of theory and speculation. It has nothing to do with plant physiology or abstruse reasoning about plant growth. It simply tells, plainly and briefly, what every one who sows a seed, makes a cutting, sets a graft, or crosses a flower wants to know. It is entirely new and original in method and matter. The cuts number 107, and are made expressly for it, direct from nature. The book treats of all kinds of cultivated plants, fruits vegetables, greenhouse plants, hardy herbs, ornamental trees, shrubs and forest trees.

CONTENTS:

I.—SEEDAGE. On Propagation by Seed.
II.—SEPARATION.
III.—LAYERAGE. Propagation by Layering.
IV.—CUTTAGE. Propagation by Cuttings.
V.—GRAFTAGE.—Including Grafting, Budding, Inarching, etc.
VI.—NURSERY LIST.—This is the great feature of the book. It is an alphabetical list of all kinds of plants, with a short statement telling which of the operations described in the first five chapters are employed in propagating them. *Over 2,000 entries* are made in the list. The following entries will give an idea of the method:

Acer (MAPLE). *Sapindaceæ.* Stocks are grown from stratified seeds, which should be sown an inch or two deep; or some species, as *A. dasycarpum,* come readily if seeds are sown as soon as ripe. Some cultural varieties are layered, but better plants are obtained by grafting. Varieties of native species are worked upon common or native stocks. The Japanese sorts are winter-worked upon imported *A. polymorphum* stocks, either by whip or veneer grafting. Maples can also be budded in summer, and they grow readily from cuttings of both ripe and soft wood.

Phyllocactus, Phyllocereus, Disocactus (LEAF CACTUS. *Cacteæ.* Fresh seeds grow readily. Sow in rather sandy soil which is well drained, and apply water as for common seeds. When the seedlings appear, remove to a light position. Cuttings from mature shoots, three to six inches in length, root readily in sharp sand. Give a temperature of about 60°, and apply only sufficient water to keep from flagging. If the cuttings are very juicy they may be laid on dry sand for several days before planting.

VII.—POLLINATION.

Price, in Library Style, cloth, wide margins, $1 Pocket Style, paper narrow margins, 50 cents.

THE MODIFICATION OF PLANTS BY CLIMATE.—By A. A. CROZIER. An essay on the influence of climate upon size, form, color, fruitfulness, etc., with a discussion on the question of acclimation. 35 pp.
Price, paper, 25 cents.

FRUIT CULTURE, and the Laying Out and Management of a Country Home.—By W. C. STRONG, Ex-President of the Massachusetts Horticultural Society, and Vice-President of the American Pomological Society. Illustrated. New revised edition, with many additions, making it the latest and freshest book on the subject.

CONTENTS:

Rural Homes—Choice of Locality—Treatment—A Good Lawn—The Approach. Fruits—Location of the Fruit Garden—Success in Fruit-Culture—Profit in Fruit-Culture. How to Procure Trees—Quality—How to Plant—Time to Plant—Preparing the Land—Fertilizers—Cutting Back—Distances for Planting. Care of the Fruit-Garden—Irrigation—Application of Fertilizers—Thinning the Fruit—Labels. The Apple—Insects Injurious to the Apple. The Pear—Dwarf Pears—Situation and Soil—Pruning—Ripening the Fruit—Insects Injurious to the Pear—Diseases. The Peach—Injurious Insects and Diseases of the Peach—Nectarines. The Plum—Insects and Diseases of the Plum—Apricots. The Cherry—Insects Injurious to the Cherry. The Quince—Insects Injurious to the Quince. The Grape—Grape-Houses—Varieties—Insects Injurious to the Grape—Mildew. The Currant—Insects Attacking the Currant—The Gooseberry. The Raspberry—The Blackberry The Strawberry. The Mulberry —The Fig—Rhubarb—Asparagus. Propagating Fruit-Trees—From the Seed—By Division—By Cuttings—By Layers—By Budding—By Grafting Insecticides—Fungicides—Recipes. Price, in one volume, 16mo., cloth, $1.

CHRYSANTHEMUM CULTURE FOR AMERICA.—By JAMES MORTON. An excellent and thorough book ; especially adapted to the culture of Chrysanthemums in America. The contents include Propagation by Grafting. Inarching and Seed. American History. Propagation by Cuttings. Exhibition Plants. Classification. Exhibition Blooms. Soil for Potting. Watering and Liquid Manure. Selection of Plants. Top-Dressing. Hints on Exhibitions. List of Synonyms. Staking and Tying General Culture. Insects and Diseases. Standard Chrysanthemums. Sports and Variations. Disbudding and Thinning. Oriental and European History. Calendar of Monthly Operations. Chrysanthemum Shows and Organizations. National Chrysanthemum Society. Early and Late-Flowering Varieties. Chrysanthemums as House-Plants—Varieties for Various Purposes. Price, cloth, $1 ; paper, 60 cents.

IMPROVING THE FARM, or Methods of Culture that shall afford a profit, and at the same time increase the fertility of the soil.—By LUCIUS D. DAVIS, of Conanicut Park Farm. The contents treat exhaustively on renewing run-down farms, and comprise the following chapters: Book-Farming. The Run-Down Farm. Will It Pay to Improve the Farm? How Farms Become Exhausted. Thorough Tillage. Rotation of Crops. Green Manuring. More About Clover. Barn-Yard Manure—How Made, Its Cost and Value. How Prepared and Applied. The Use of Wood-Ashes. Commercial Fertilizers. Special Fertilizers. Complete Manures. Experiments with Fertilizers. Stock on the Farm. Providing Food for Stock. Specialties in Farming. Price, cloth, $1.

LANDSCAPE-GARDENING.—By ELIAS A. LONG. A practical treatise comprising 32 diagrams of actual grounds and parts of grounds, with copious explanations. Of the diagrams, all but nine have appeared in the serial, "Taste and Tact in Arranging Ornamental Grounds," which has been so attractive a feature of *Popular Gardening* and *American Gardening* during the past year. But in the new form the matter has been entirely rewritten. Printed on heavy plate paper, it is unsurpassed for beauty by any other work on Landscape Gardening.

Price, 50 cents.

THE BUSINESS HEN.—Breeding and Feeding Poultry for Profit. The pat title of a unique book is The Business Hen. A condensed and, practical little enclycopedia of profitable poultry-keeping. P. H. Jacobs, Henry Hale, James Rankin, J. H. Drevenstedt and others equally well known have written chapters on their specialties, the whole being skillfully arranged and carefully edited by H. W. Collingwood, managing editor of *The Rural New-Yorker*. Starting with the question, "What is an Egg?" the book goes on step by step to indicate the most favorable conditions for developing the egg into a "Business Hen." Incubation, care of chicks, treatment of diseases, selection and breeding, feeding and housing, are all discussed in a clear and simple manner. Two successful egg-farms are described in detail. On one of these farms the owner has succeeded in developing a flock of 600 hens that average over 200 eggs each per year.

Price, cloth, 75 cents; paper, 40 cents.

FIRST LESSONS IN AGRICULTURE. (*2nd Edition, Revised and Enlarged.*)—By F. A. Gulley, M. S., Professor of Agriculture in the Agricultural College of Mississippi. This book discusses the more important principles which underlie agriculture in a plain, simple way, within the comprehension of students and readers who have not studied chemistry, botany, and other branches of science related to agriculture. It supplies a much-needed text-book for common schools, and is useful for the practical farmer. Includes all the latest developments in agricultural science as applied to the subject.

Price, cloth, $1. Special prices for Schools and Colleges.

THE NEW POTATO CULTURE.—By ELBERT S. CARMAN. This book gives the result of 15 years' experiment work on The Rural ground. It treats particularly of: How to increase the crop without corresponding cost of production. Manures and fertilizers: kinds and methods of application. The soil, and how to put it in right condition. Depth of planting. How much seed to plant. Methods of culture. The Rural trench system. Varieties, etc., etc.

Nothing old or worn-out about this book. It treats of new and profitable methods; in fact, of *The NEW Potato Culture*. It is respectfully submitted that these experiments at The Rural grounds have, directly and indirectly, thrown more light upon the various problems involved in successful potato-culture than any other experiments that have been carried on in America.

Price, cloth, 75 cents; paper, 40 cents.

HORTICULTURIST'S RULE-BOOK.—By Professor L. H. BAILEY, Editor of *American Gardening*, Horticulturist of the Cornell Experiment Station, and Professor of Horticulture in Cornell University. It contains in handy and concise form, a great number of Rules and Recipes required by gardeners, fruit-growers, truckers, florists, farmers, etc.

Synopsis of Contents: Injurious insects, with preventives and remedies. Fungicides for plant diseases. Plant diseases, with preventives and remedies. Injuries from mice, rabbits, birds, etc., with preventives and remedies. Waxes and washes for grafting and for wounds. Cements, paints, etc. *Seed Tables:* Quantities required for sowing given areas. Weight and size of seeds. Longevity of seeds. Time required for seeds to germinate. *Planting Tables:* Dates for sowing seeds in different latitudes. Tender and hardy vegetables. Distances apart for planting. *Maturity and Yields:* Time required for maturity of vegetables; for bearing of fruit plants. Average yields of crops. Keeping and storing fruits and vegetables. *Propagation of Plants:* Ways of grafting and budding. Methods by which fruits are propagated. Stocks used for fruits. *Standard Measures and Sizes:* Standard flower-pots. Standard and legal measures. English measures for sale of fruits and vegetables. Quantities of water held in pipes and tanks. Effect of wind in cooling off glass roofs. Per cent. of light reflected from glass at various angles of inclination. Weights of various varieties of apples per bushel. Amount of various products yielded by given quantities of fruit. Labels. Loudon's rules of horticulture. Rules of nomenclature. Rules for exhibition. Weather signs and protection from frost. *Collecting and Preserving:* How to make an herbarium. Preserving and printing of flowers and other parts of plants. Keeping cut-flowers. How to collect and preserve insects. Chemical composition of fruits and vegetables, and seeds fertilizers, soils and vegetables. *Names and Histories:* Vegetables which have different names in England and America. Derivation of names of various fruits and vegetables. Names of fruits and vegetables in various languages. Glossary. Calendar.

Price, cloth, $1; paper, 60 cents.

CROSS-BREEDING AND HYBRIDIZING:—The Philosophy of the Crossing of Plants considered with reference to their Cultivation—How to Improve plants by Hybridizing.--By L. H. BAILEY. It is the only book accessible to American horticulture which gives the reasons, discouragements, possibilities and limitations of Cross-Breeding. Every man who owns a plant should have it, if for no other reason than to post himself upon one of the leading practices of the day. The pamphlet contains also a bibliography of the subject, including over 400 entries.

Price, paper, 40 cents.

CHEMICALS AND CLOVER.—By H. W. COLLINGWOOD, Managing Editor of *The Rural New-Yorker*. A concise and practical discussion of the all-important topic of commercial fertilizers in connection with green manuring in bringing up worn-out soils, and in general farm practice.

Price, paper, 20 cents.

ANNALS OF HORTICULTURE, Vol. IV.—Bright, New, Clean and Fresh. These Annals are entirely rewritten every year. They are the *only records* of the progress in horticulture. Exhaustive lists of all the plants introduced in 1892, with descriptions, directories, full accounts of all new discoveries, new tools, and a wealth of practical matter for *Gardeners, Fruit-Growers, Florists, Vegetable-Gardeners* and *Landscape-Gardeners*, comprise its contents.

Ready soon. Illustrated. Vol. IV., cloth $1. Vols. I., II and III. at the same price.

INSECTS AND INSECTICIDES.—A practical Manual concerning Noxious Insects and the Methods of Preventing their Injuries. By CLARENCE M. WEED, Professor of Entomology and Zoölogy, New Hampshire State College.

I think that you have gotten together a very useful and valuable little book.—Dr. C. V. RILEY, *U. S. Entomologist.*

It is excellent. I must congratulate you on the skill you have displayed in putting in the most important insects, and the complete manner in which you have done the work.—JAMES FLETCHER, *Dominion Entomologist.*

I am well pleased with it. There is certainly a demand for just such a work.—Dr. F. M. HEXAMER, *Editor American Agriculturist.*

Price, cloth, $1.25.

THE CAULIFLOWER.—By A. A. CROZIER. Teacher and Practical Origin and History of this increasingly important and always delicious vegetable.

The Cauliflower Industry.—In Europe. In the United States. Importation of Cauliflowers.

Management of the Crop.—Soil. Fertilizers. Planting. Cultivating, Harvesting Keeping. Marketing.

The Early Crop.—Caution against planting it largely. Special directions. Buttoning.

Cauliflower Regions of the United States.—Upper Atlantic Coast. Lake Region. Prairie Region. Cauliflowers in the South. The Pacific Coast.

Insect and Fungous Enemies.—Flea-beetle. Cut-worms. Cabbage-maggot. Cabbage-worm. Stem-rot. Damping-off. Black-leg.

Cauliflower Seed.—Importance of careful selection. Where the seed is grown. Influence of climate. American-grown seed.

Varieties.—Descriptive catalogue. Order of earliness. Variety tests. Best varieties.

Broccoli.—Difference between Broccoli and Cauliflower. Cultivation, use and varieties of Broccoli.

Cooking Cauliflower.—Digestibility. Nutritive value. Chemical composition. Recipes.

Price, cloth, $1.

PRACTICAL FARM CHEMISTRY.—A Practical Handbook of Profitable Crop-Feeding, written for Practical Men. By T. GREINER.
Part I. The Raw Materials of Plant-Food.
Part II. The Available Sources of Supply.
Part III. Principles of Economic Application, or Manuring for Money.

This work, written in plainest language, is intended to assist the farmer in the selection, purchase and application of plant-foods. If you wish to learn ways how to save money in procuring manurial substances, and how to make money by their proper use, read this book. If you want your boy to learn the principle of crop-feeding, and become a successful farmer, give him a copy of this book. The cost of the book will be returned a hundred-fold to every reader who peruses its pages with care and applies its teachings to practice.

Price, cloth, $1.

SPRAYING CROPS.—Why, When and How to Do It.—By PROF. CLARENCE M. WEED. A handy volume of about 100 pages; illustrated. Covers the whole field of the insect and fungous enemies of crops for which the spray is used. The following topics are discussed in a concise, practical manner:

Spraying Against Insects. Feeding Habits of Insects. Spraying Against Fungous Diseases. The Philosophy of Spraying. Spraying Apparatus. Spraying Trees in Blossom. Precautions in Spraying. Insecticides used in Spraying. Fungicides used in Spraying. Combining Insecticides and Fungicides. Cost of Spraying Materials. Prejudice Against Spraying. Spraying the Larger Fruits. Spraying Small Fruits and Nursery Stock. Spraying Shade Trees, Ornamental Plants and Flowers. Spraying Vegetables, Field Crops and Domestic Animals.

Price in stiff paper cover, 50 cents; flexible cloth, 75 cents.

www.ingramcontent.com/pod-product-compliance
Lightning Source LLC
Chambersburg PA
CBHW020153170426
43199CB00010B/1023